SOLUTIONS MANUAL TO
INSTRUMENTATION FOR ENGINEERING MEASUREMENTS

Second Edition

JAMES W. DALLY
University of Maryland

WILLIAM F. RILEY
KENNETH G. McCONNELL
Iowa State University

JOHN WILEY & SONS, INC.
New York • Chichester • Brisbane • Toronto • Singapore

Copyright © 1993 by John Wiley & Sons, Inc.

This material may be reproduced for testing or instructional purposes by people using the text.

ISBN 0-471-58311-1

Printed in the United States of America

10 9 8 7 6 5 4 3 2 1

CONTENTS

CHAPTER 1 1
CHAPTER 2 12
CHAPTER 3 27
CHAPTER 4 49
CHAPTER 5 70
CHAPTER 6 88
CHAPTER 7 113
CHAPTER 8 141
CHAPTER 9 159
CHAPTER 10 183
CHAPTER 11 203
CHAPTER 12 234
CHAPTER 13 259

ENGINEERING MEASUREMENTS by J. W. DALLY, W. F. RILEY, AND K. G. McCONNELL

Exercise 1.1

Electronic systems provide data that more accurately and completely characterize the design or process being evaluated. Also, the electronic system provides an electrical output signal that can be used directly for analog control of processes or digitized for automatic data reduction. Finally, the sensors associated with electronic sensors are small and do not affect the process being evaluated.

Exercise 1.2

Six to eight subsystems are used in most electronic instrumentation systems.

1. Transducer: An analog device that converts a change in the thermal or mechanical quantity being sensed into an electrical quantity.

2. Power supply: Provides the energy required to drive the transducer.

3. Signal conditioners: Circuits that convert, compensate or manipulate the transducer output into a more usable electrical quantity.

4. Amplifiers: Increase the voltage from the transducer-signal conditioner.

5. Recorders: Both analog or digital recorders are used to display or store signals.

6. Data processors: Micro computers that accept a digitized signal and perform computations in accordance with programmed instructions.

7. Command generator: Provides a control signal that represents the profile of an important parameter in a process.

8. Process controllers: Monitors and adjusts process variables to maintain accurate control of a given process.

Exercise 1.3

Applications of electronic instrumentation systems include:

1. Engineering analyses of machine components, structures, and vehicles.

2. Process monitoring for open-loop control.

3. Process monitoring for closed-loop control.

ENGINEERING MEASUREMENTS by J. W. DALLY, W. F. RILEY, AND K. G. McCONNELL

Exercise 1.4

Engineering analyses are conducted to test and evaluate new or modified products before they are released for public use. The preferred approach utilizes a mathematic analysis followed by sufficient experimental testing to ensure safe and reliable use of the product under development. In the mathematical analysis, an analytical model of the product is formulated that describes the operation of the product under all conceivable operating conditions. The behavior of the product is studied by employing a computer to numerically exercise the analytical model. The results provide design engineers with an indication of the adequacy of the design and an estimate of the performance of the product under service conditions. A prototype or scale model of the product is fabricated for use in the experimental phase of the program. Instruments are placed on the prototype (or model) and measurements are made to test the strength, performance, and reliability of the product. When the results of the mathematical analysis and the experiments with the prototype are in agreement, the design engineers can be confident that the product will perform in accordance with the design and that the customer will be satisfied with the performance, quality, safety, and reliability of the product.

Exercise 1.5

The preferred approach in performing an engineering analysis employs a combination of theoretical and experimental methods. The theoretical analysis is conducted to ensure a thorough understanding of the product and its performance. The significance of the results of the theoretical analysis should be completely evaluated and any shortcomings of the analysis should be clearly identified. An experiment is then designed to verify the analytical model to insure the adequacy of the numerical methods used in generating the analytical results.

Exercise 1.6

The two types of process control are open loop (monitoring control) or closed loop (automatic control).

In open-loop control, an operator monitors a process by observing the output from several transducers. The operator then adjusts process parameters to correct for deviations and to bring the process within control bands.

In closed-loop control, the operator is eliminated. The signals from the transducers are compared to command signals that define the process parameters. Differences between these two signals represent control errors. The "error" signals are amplified and used to drive devices that correct the process.

Exercise 1.7

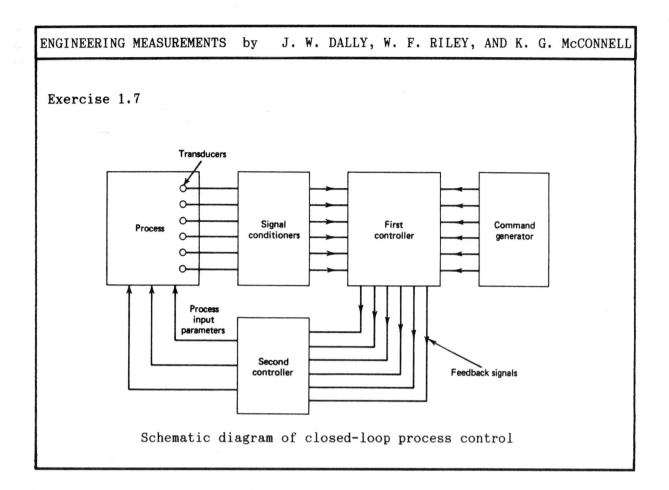

Schematic diagram of closed-loop process control

Exercise 1.8

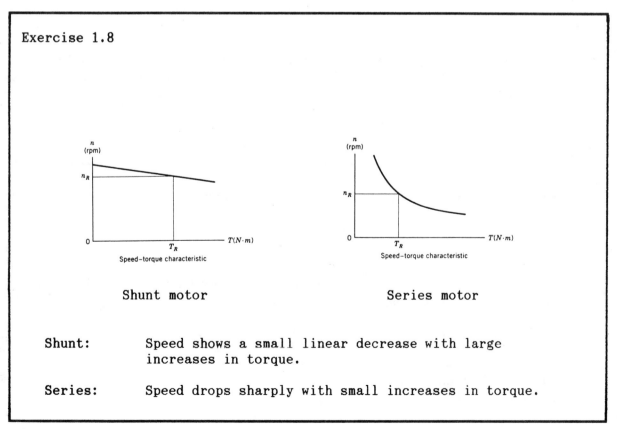

Shunt motor   Series motor

Shunt: Speed shows a small linear decrease with large increases in torque.

Series: Speed drops sharply with small increases in torque.

Exercise 1.9

Stepping motors differ from traditional dc motors. The differences are:

1. The stator incorporates a large number of small field poles deployed around the circumference.

2. Each pole piece is made with several teeth.

3. The armature resembles a gear with teeth uniformly spaced about the periphery.

4. The armature core is fabricated with permanent magnets that eliminate the need for armature coils.

5. The motor is driven by a train of pulses delivered to the field coils.

6. The pulse characteristics are matched to the motor inertia to give a rotation of one step (one tooth) for each pulse.

Exercise 1.10

A dc motor is used in closed-loop control where the number of rotations is monitored and feedback signals from position transducers control the angular rotation.

The rotation of a stepping motor is an open-loop process is controlled by the number of pulses. If a specific number of rotations is needed then the pulse count is determined and applied.

DC motors are available in a wide range of torque capacities. Stepping motors are used for lower torque applications.

Exercise 1.11

Flow control devices include:

Solenoid valves: Solenoid valves are either open or shut; therefore, they are used only for bang-bang control. Low cost.

Motorized valves: Flow rate is accurately controlled but it exhibits relatively low frequency response. Moderate cost.

Servo control valves: Flow rate is accurately controlled with very high frequency response. Very expensive.

ENGINEERING MEASUREMENTS by J. W. DALLY, W. F. RILEY, AND K. G. McCONNELL

Exercise 1.12

A 50-tooth pitch, four-pole, single-phase stator gives 200 steps per revolution or $360°/200 = 1.8°$ per pulse.

$$\Delta\ell = \left(\frac{1.8}{360}\right)(1) = 0.005 \text{ mm} \quad \text{or} \quad \Delta\ell = \frac{1}{200}(1) = 0.005 \text{ mm}$$

Exercise 1.13

The advantages include:

1. Lead-screw positioning devices are precise with submicrometer accuracy.
2. They provide accurate control over a wide range if long lead screws are used.

The disadvantages include:

1. Backlash (clearance between thread and nut) when the direction of rotation is changed.
2. The velocity of the drive nut is slow; 100 mm/s is near the upper limit.

Exercise 1.14

$$A = \pi d^2/4 = \pi(4)^2/4 = 4\pi \text{ in.}^2$$

$$V = AL = 4\pi L \text{ in.}^3$$

$$\frac{dV}{dt} = A\frac{dL}{dt} \text{ in.}^3/\text{min}$$

$$v = \frac{dL}{dt} = \frac{dV/dt}{A} = \frac{10(231)}{4\pi} = 183.8 \text{ in./min.}$$

Exercise 1.15

$$F = pA = (3000 - 400)[\pi(4)^2/4] = 32{,}672 \text{ lbs.}$$

Exercise 1.16

Immersion heaters: Resistive element protected by sheathing so that it can be immersed in fluids without danger of shorting.

Coils of nichrome wire: Coils are not protected by sheathing; therefore, they are usually used only in low temperature ovens that use air convection for heat transfer.

Glow bars, Quartz lamps: Infrared heating elements. Heat is transferred by radiation. They are used when high heating rates ($\Delta T/\Delta t$) are required or at higher temperatures.

Exercise 1.17

Sources of error that must be included in the design of an instrument system include:

1. Accumulation of accepted error in each element.

2. Improper functioning of any element.

3. Effect of the transducer on the process.

4. Dual sensitivity error.

5. Electronic noise.

6. Operator error.

Exercise 1.18

$$d = \pm 0.01(200) = \pm 2.0 \text{ mV.}$$

% FS	Reading	Deviation	% Error
100	200	2.0	1.00
75	150	2.0	1.33
50	100	2.0	2.00
25	50	2.0	4.00
12.5	25	2.0	8.00

Conclusion: Use instruments near full scale when high accuracy is required.

Exercise 1.19

From Eq. (1-5):

Case 1:
$$\mathcal{E} = \sqrt{(0.05)^2 + (0.01)^2 + (0.01)^2 + (0.01)^2 + (0.01)^2}$$
$$= 0.0539 = 5.39 \%$$

Case 2:
$$\mathcal{E} = \sqrt{(0.01)^2 + (0.01)^2 + (0.02)^2 + (0.02)^2 + (0.03)^2}$$
$$= 0.0436 = 4.36 \%$$

Case 3:
$$\mathcal{E} = \sqrt{(0.01)^2 + (0.01)^2 + (0.01)^2 + (0.01)^2 + (0.01)^2}$$
$$= 0.0224 = 2.24 \%$$

Case 4:
$$\mathcal{E} = \sqrt{(0.02)^2 + (0.02)^2 + (0.05)^2 + (0.02)^2 + (0.02)^2}$$
$$= 0.0640 = 6.40 \%$$

Exercise 1.20

Range: Maximum input for a specified deviation from linear response.

Span: Difference between upper and lower limits of operation.

Exercise 1.21

From Eq. (1.7):
$$Q_o = SQ_i + Z_0 = SQ_{ia}$$

where
Q_i = true input

Q_{ia} = apparent input

$$\%\mathcal{E} = \frac{Q_{ia} - Q_i}{Q_i}(100) = \frac{Q_o/S - (Q_o - Z_0)/S}{(Q_o - Z_0)/S}(100) = \frac{Z_0}{Q_o - Z_0}(100)$$

ENGINEERING MEASUREMENTS by J. W. DALLY, W. F. RILEY, AND K. G. McCONNELL

Exercise 1.22

From Eq. (1.8): $\quad Q_o = S_1 Q_i = S Q_{ia}$

where $\quad Q_i$ = true input

$\quad Q_{ia}$ = apparent input

$$\%\mathcal{E} = \frac{Q_{ia} - Q_i}{Q_i}(100) = \frac{Q_o/S - Q_o/S_1}{Q_o/S_1}(100) = \frac{S_1 - S}{S}(100)$$

Exercise 1.23

From Eq. (1.7): $\quad Q_o = S_1 Q_i + Z_0 = S Q_{ia}$

where $\quad Q_i$ = true input

$\quad Q_{ia}$ = apparent input

$$\%\mathcal{E} = \frac{Q_{ia} - Q_i}{Q_i}(100) = \frac{Q_o/S - (Q_o - Z_0)/S_1}{(Q_o - Z_0)/S_1}(100) = \left[\frac{Q_o S_1}{(Q_o - Z_0)S} - 1\right](100)$$

Exercise 1.24

1. Large Accelerometer on a vibrating member.

2. Orifice or flow nozzle in a pipe to measure the flow.

3. Dial gage measuring deflections of a thin rubber membrane.

Exercise 1.25

1. Strain gages respond to both strain and temperature change. Response due to temperature change during a test can be interpreted as a load induced strain.

2. Force gages may respond to both force and acceleration.

ENGINEERING MEASUREMENTS by J. W. DALLY, W. F. RILEY, AND K. G. McCONNELL

Exercise 1.26

(a) Over long periods of time very small changes in a number of different parameters accumulate to produce very large errors. Drift of amplifiers and recording instruments are perhaps the most common source of error.

(b) If the instruments can be rezeroed periodically, the drift error can be eliminated. Errors due to temperature change can be eliminated by placing the instruments in a temperature controlled room and by using transducers with temperature compensation. Errors due to humidity and other environmental effects can be eliminated by waterproofing the transducers and employing hermetically sealed electronics.

Exercise 1.27

Let A = Full scale output of the amplifier

Drift = $0.005(A)(6) = 0.03A$

$Q_{oA} = 0.50A + 0.03A = 0.53A$

$Q_{oT} = 0.50A$

$\%\mathcal{E} = \dfrac{Q_{oA} - Q_{oT}}{Q_{oT}}(100) = \dfrac{0.53A - 0.50A}{0.50A}(100)$

$= 0.060(100) = 6.0\ \%$

Exercise 1.28

$Q_{oT} = 5.0(200) = 1000$

$Q_{oA} = 5.0(200) + 0.05(30) = 1001.5$

$\%\mathcal{E} = \dfrac{Q_{oA} - Q_{oT}}{Q_{oT}}(100) = \dfrac{1001.5 - 1000}{1000}(100)$

$= 0.00150(100) = 0.150\ \%$

Exercise 1.29

$$S = \frac{\Delta R/R}{\varepsilon}$$

With ΔR and ε fixed:

$$\frac{\Delta R}{\varepsilon} = R_o S_o = R_L S_L$$

$$S_L = \frac{R_o S_o}{R_L} = \frac{350(2.00)}{350 + 4} = 1.977$$

Exercise 1.30

$$\Delta R = R\, \gamma\, \Delta T = 4(0.0039)(20) = 0.312\ \Omega$$

$$\varepsilon_{apparent} = \frac{\Delta R/R}{S} = \frac{0.312/354}{1.977} = 0.000446 = 446\ \mu m/m$$

Exercise 1.31

Other sources of error include:

Lead wire effects: Long lead wires affect the sensitivity of resistive sensors.

Electronic noise: Fluctuating electrostatic and magnetic fields generate noise signals in lead wires that are superimposed on the signal representing the measurement.

Operator errors: Calibration errors, reading errors, and poor placement of transducers are examples of operator error.

Exercise 1.32

For aluminum:
$$\rho = 2800\ kg/m^3$$

$$m_p = \rho V = 2800(\pi/4)(0.200)^2(0.0007)$$

$$= 0.06158\ kg = 61.58\ grams$$

For $m_a = 0.05 m_p = 0.05(61.58) = 3.08$ grams: $\mathscr{E} = 13.3\%$

For $m_a = 0.10 m_p = 0.10(61.58) = 6.16$ grams: $\mathscr{E} = 25.9\%$

ENGINEERING MEASUREMENTS by J. W. DALLY, W. F. RILEY, AND K. G. McCONNELL

Exercise 1.33

Encourage outside reading of textbooks and journals.

(a) Check output voltage (or current) as a function of load on the power supply.

(b) Special attention to the effect of the transducer on the measurement.

(c) Special attention to linear range of output for the sensor being used.

(d) Special attention to the problem of stability.

(e) Special attention to problems of circuit loading by the voltmeter.

Exercise 1.34

(a) Mount a typical gage on a cantilever beam. The free end of the beam can be deflection limited to produce a desired level of strain.

(b) A static pressure can be produced by using a hydraulic ram and dead weights.

(c) Freezing and boiling points of water (corrected for atmospheric pressure) provide convenient calibration temperatures.

(d) A micrometer can be used to produce accurate displacements.

(e) Accelerations of 1 g can be obtained by rotating the accelerometer in the earth's gravitational field.

Exercise 1.35

Electronic noise usually results from spurious signals that are picked up by the lead wires. Noise can be minimized with proper shielding. In some measurements with a very small signal, noise from a properly shielded lead-wire installation may still be objectionable. In these cases, notch filters that block passage of a narrow band of frequencies can be used to eliminate most of the noise, since it usually exhibits the 60-Hz power-line frequency.

ENGINEERING MEASUREMENTS by J. W. DALLY, W. F. RILEY, AND K. G. McCONNELL

Exercise 2.1

(a)	Name	Symbol	Unit	Abbreviation
	Force	f	newton	N
	Charge	q	coulomb	C
	Electric field strength	$\mathcal{E}$	volt/meter	V/m
(b)	Energy	w	joule	J
	Current	i	ampere	A
	Magnetic flux density	$\mathcal{B}$	tesla	T
(c)	Power	p	watt	W
	Voltage	v	volt	V
	Magnetic Flux	ϕ	weber	Wb

Exercise 2.2

$$p = vi = v\left(\frac{v}{R_0}\right) = \frac{v^2}{R_0}$$

Voltage	Power p (W)			
	$R_0 = 50\ k\Omega$	$R_0 = 100\ k\Omega$	$R_0 = 250\ k\Omega$	$R_0 = 500\ k\Omega$
0	0	0	0	0
200	0.8	0.4	0.16	0.08
400	3.2	1.6	0.64	0.32
600	7.2	3.6	1.44	0.72
800	12.8	6.4	2.56	1.28
1000	20.0	10.0	4.00	2.00

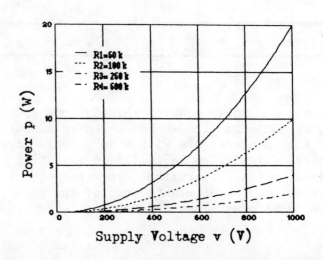

Exercise 2.3

From Eqs. 2.3, 2.11, and 2.13:

$$v_o = Ri + \frac{q}{C} = R\frac{dq}{dt} + \frac{q}{C}$$

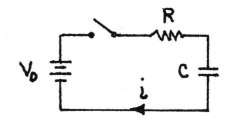

Solving for q:

$$q = v_o C(1 - e^{-t/RC})$$

and

$$i = \frac{dq}{dt} = v_o C \left(\frac{e^{-t/RC}}{RC}\right)$$

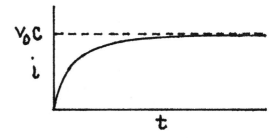

As R becomes smaller and smaller, the charge approaches a step function and the current approaches a Dirac delta function.

Exercise 2.4

From Exercise 2.3: $\quad i = \frac{v_o}{R} e^{-t/RC}$

Case	v_o (V)	R (Ω)	C (F)	RC (s)	i (A)
(a)	10	10	$10(10^{-6})$	$100(10^{-6})$	1.0
(b)	5	10	$200(10^{-12})$	$2(10^{-9})$	0.5
(c)	100	10	1	10	10.0
(d)	3	10	$1500(10^{-12})$	$15(10^{-9})$	0.3

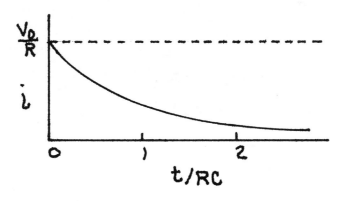

Exercise 2.5

From Eqs. 2.13, and 2.14:

$$q = Cv$$

$$w = \frac{1}{2} Cv^2$$

Case	v_0 (V)	C (F)	q (C)	w (J)
(a)	10	$10(10^{-6})$	$100(10^{-6})$	$500(10^{-6})$
(b)	5	$200(10^{-12})$	$1000(10^{-12})$	$2.5(10^{-9})$
(c)	100	1	100	5000
(d)	3	$1500(10^{-12})$	$4500(10^{-12})$	$6.75(10^{-9})$

Exercise 2.6

$$v(t) = a \sin 2\pi ft = a \sin \omega t$$

From Eq. 2.16:

$$L \frac{di}{dt} = v(t) = a \sin \omega t$$

$$i(0) = 0$$

$$\int_0^i di = \frac{a}{L} \int_0^t \sin(\omega t)\, dt$$

which yields:

$$i = \frac{a}{L\omega}(1 - \cos \omega t)$$

Note: This solution is valid only if R = 0.

Exercise 2.7

(a) From Exercise 2.6:
$$v(t) = a \sin 2\pi ft = a \sin \omega t$$
$$i = \frac{a}{L\omega}(1 - \cos \omega t)$$

(1) For $f = 60$ Hz, $L = 10$ mH, and $a = 10$ V:
$$\frac{a}{L\omega} = \frac{10}{0.010(2\pi)(60)} = 2.65 \text{ A}$$

(2) For $f = 1$ MHz, $L = 10$ μH, and $a = 5$ V:
$$\frac{a}{L\omega} = \frac{5}{1(10^6)(2\pi)(10)(10^{-6})} = 79.6(10^{-3}) \text{ A} = 79.6 \text{ mA}$$

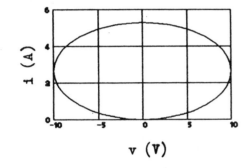 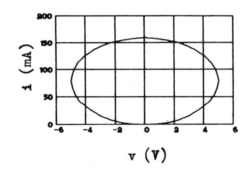

Case 1   Case 2

(b) The plot is an ellipse or circle (depends on scaling) with an offset.

(c) The plot is stable if $R = 0$.

Exercise 2.8

From Eq. 2.17:
$$w = \frac{1}{2} Li^2$$

(1) For $f = 60$ Hz, $L = 10$ mH, and $a = 10$ V:
$$w = \frac{1}{2}(10)(10^{-3})(2.65)^2 = 0.0351 \text{ J}$$

(2) For $f = 1$ MHz, $L = 10$ μH, and $a = 5$ V:
$$w = \frac{1}{2}(10)(10^{-6})[(79.6)(10^{-3})]^2 = 31.7(10^{-9}) \text{ J}$$

Exercise 2.9

Kirchhoff's First law:

The algebraic sum of currents entering and leaving a junction is zero. This law simply states that charge is neither created nor destroyed at the junction and it does not accumulate at this point. This law accounts for the flow of electrons at a junction in the same way that continuity accounts for the mass of fluid at a pipe junction.

Exercise 2.10

Kirchhoff's Second Law:

The algebraic sum of the changes in potential around a closed loop is zero at each instant. This law is a simple statement of the conservation of energy. The potential energy given a charge by the voltage source is lost in going through the resistors.

Exercise 2.11

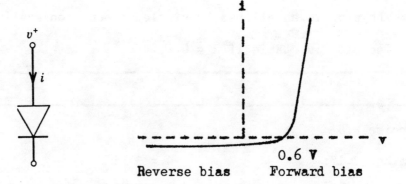

Ideally, a diode is a one way switch that is closed (allows current to flow) for positive bias voltages and is open (no current flows) for negative bias voltages. In real diodes, a positive threshold bias of 0.6 V is required before significant current flow begins while a very small current flow exists for a negative bias as shown in sketch (b).

Exercise 2.12

For $v_1 = 10 \sin(120\pi t)$:

$v_2 = 10 \sin(120\pi t)$ for positive values of $\sin(120\pi t)$

$v_2 = 0$ for negative values of $\sin(120\pi t)$

This response yields the rectified output voltage shown in the sketch below.

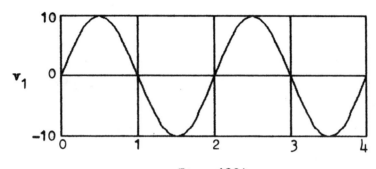

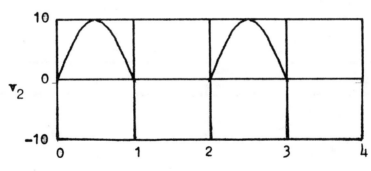

Exercise 2.13

The two primary applications for transistors are:

1. Amplifiers
2. Electronic switches

Exercise 2.14

For $V_{CC} = 10$ V and $R_L = 400$ Ω:

$$V_{CC}/R_L = 10/400 = 0.025 = 25 \text{ mA}$$

Draw a load line on Fig. 2.5b by using $V_{CC}/R_L = 25$ mA and $V_{CC} = 10$ V.

From this load line for $i_i = [0.2 - 0.1 \sin(\omega t)](10^{-3})$ A:

$$i_c(\max) = 15 \text{ mA} \quad \text{and} \quad i_c(\min) = 5 \text{ mA}$$

Exercise 2.15

When the input is low, the output voltage v_{CE} is high since no current flows through the transistor. When the input goes high, the output voltage v_{CE} is low since the transistor permits current to flow. Thus, the transistor operates as an electronic switch.

Exercise 2.16

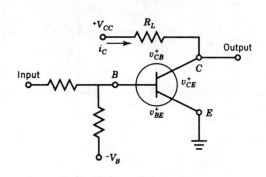

When the switch is open:

$$i_{co} \cong 1 \text{ ma}$$

When the switch is closed:

$$i_{cc} \cong 26 \text{ ma}$$

Thus, for the electronic switch:

$$i_{cc}/i_{co} \cong 26$$

For a mechanical switch

$$i_{cc}/i_{co} \cong \infty$$

Any resistance R across terminals CE of the electronic switch can further alter its operating characteristics.

Exercise 2.17

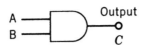

AND gate		
A	B	C
0	0	0
0	1	0
1	0	0
1	1	1

OR gate		
A	B	C
0	0	0
1	0	1
0	1	1
1	1	1

Exercise 2.18

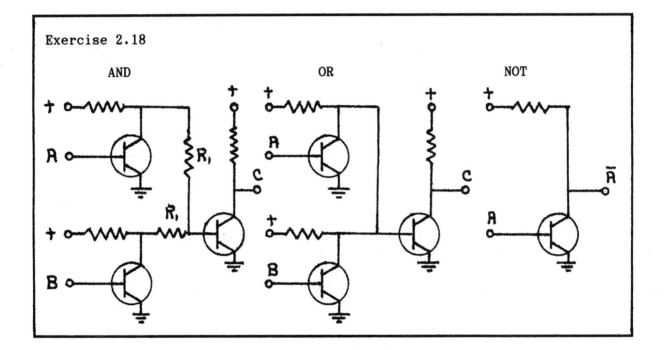

Exercise 2.19

For resistors in series: $\quad v_s = iR_1 + iR_2 = i(R_1 + R_2) = iR_e$

Therefore, $\quad R_e = R_1 + R_2$

For resistors in parallel: $\quad i = i_1 + i_2 + i_3 = \dfrac{v_s}{R_1} + \dfrac{v_s}{R_2} + \dfrac{v_s}{R_3} = \dfrac{v_s}{R_e}$

Therefore, $\quad \dfrac{1}{R_e} = \dfrac{1}{R_1} + \dfrac{1}{R_2} + \dfrac{1}{R_3}$

ENGINEERING MEASUREMENTS by J. W. DALLY, W. F. RILEY, AND K. G. McCONNELL

Exercise 2.20

$$x = A_0 \cos \omega t$$
$$\dot{x} = -\omega A_0 \sin \omega t = \omega A_0 \cos(\omega t + \pi/2)$$
$$\ddot{x} = -\omega^2 A_0 \cos \omega t = \omega^2 A_0 \cos(\omega t + \pi)$$

Exercise 2.21

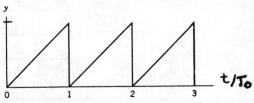

$$y(t) = \frac{a_0}{2} + \sum_{n=1}^{\infty} a_n \cos(n\omega_0 t) + b_n \sin(n\omega_0 t)$$

Let: $\quad f(t) = \dfrac{At}{T_0} \qquad \omega_0 T_0 = 2\pi \qquad z = n\omega_0 t \qquad dz = n\omega_0\, dt$

$$a_0 = \frac{2}{T_0} \int_0^{T_0} f(t)\, dt = \frac{2A}{T_0^2} \int_0^{T_0} t\, dt = \frac{2A}{T_0^2}\left[\frac{t^2}{2}\right]_0^{T_0} = A$$

$$a_n = \frac{2}{T_0} \int_0^{T_0} \frac{At}{T_0} \cos(n\omega_0 t)\, dt = \frac{2A}{T_0^2 (n\omega_0)^2} \int_0^{2n\pi} z \cos z\, dz$$

$$= \frac{2A}{(2n\pi)^2}\Big[\cos z + z \sin z\Big]_0^{2n\pi} = 0$$

$$b_n = \frac{2A}{T_0^2} \int_0^{T_0} t \sin(n\omega_0 t)\, dt = \frac{2A}{T_0^2 (n\omega_0)^2} \int_0^{2n\pi} z \sin z\, dz$$

$$= \frac{2A}{(2n\pi)^2}\Big[\sin z + z \cos z\Big]_0^{2n\pi} = -\frac{A}{n\pi}$$

$$y(t) = A\left[\frac{1}{2} - \frac{1}{\pi}\sum_{n=1}^{\infty} \frac{1}{n} \sin(n\omega_0 t)\right]$$

Exercise 2.22

(a) $\quad Q = A_0 e^{j(\omega t - \phi)} \qquad \dot{Q} = j\omega A_0 e^{j(\omega t - \phi)} \qquad \ddot{Q} = \omega^2 A_0 e^{j(\omega t - \phi)}$

(b) The phase angle ϕ affects each quantity by the same amount. All three are shifted ϕ radians with respect to a reference phasor $A_0 e^{j\omega t}$.

Exercise 2.23

$$Z = R + j(L\omega - \frac{1}{C\omega})$$

From Eqs. 2.38 and 2.39:

$$|Z| = \sqrt{(Re)^2 + (Im)^2} = \sqrt{(R)^2 + (L\omega - 1/C\omega)^2}$$

$$\phi = \tan^{-1}\frac{(Im)}{(Re)} = \tan^{-1}\frac{L\omega - 1/C\omega}{R}$$

Exercise 2.24

(a) From Eqs. 2.13, 2.15, and 2.19:

$$v_s(t) = Ri + \frac{q}{C} \qquad v_o = \frac{q}{C} \qquad i = \dot{q} = C\dot{v}_o$$

Thus: $\qquad RC\dot{v}_o + v_o = v_s(t) = v_i e^{j\omega t}$

Assume: $\qquad v_o(t) = v_o e^{j\omega t} \qquad \dot{v}_o(t) = j\omega v_o e^{j\omega t}$

Then: $\qquad (jRC\omega + 1)v_o e^{j\omega t} = v_i e^{j\omega t}$

$$\frac{v_o}{v_i} = \left[\frac{1}{1 + jRC\omega}\right] = \frac{1}{\sqrt{1 + (RC\omega)^2}} e^{-j\phi}$$

$$\tan\phi = RC\omega$$

(b)

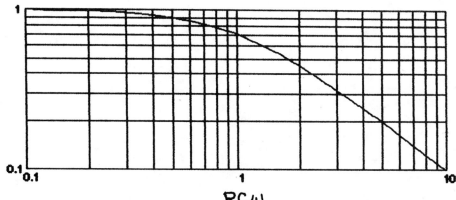

(c) $\qquad Z_C = \frac{1}{jC\omega}$

Thus, Z_C becomes very small for large values of ω.

Exercise 2.25

For the circuit shown in Fig. E2.25:

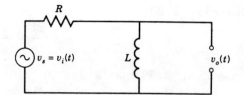

$$v_s = v_i(t) = Ri + Z_L i$$

Assume:

$$v_i(t) = v_i e^{j\omega t}$$

$$i = \frac{v_i}{R + Z_L} e^{j\omega t} = \frac{v_i}{R + j\omega L} e^{j\omega t} = \frac{R - j\omega L}{R^2 + (\omega L)^2} v_i e^{j\omega t}$$

From Eq. 2.38:

$$i_0 = \sqrt{(Re)^2 + (Im)^2}$$
$$= \frac{[R^2 + (-\omega L)^2]^{1/2}}{R^2 + (\omega L)^2} v_i = \frac{1}{[R^2 + (\omega L)^2]^{1/2}} v_i$$

$$\phi_1 = \tan^{-1} \frac{Im}{Re} = \frac{-\omega L}{R}$$

$$i = i_0 e^{j(\omega t - \phi_1)} = \frac{1}{[R^2 + (\omega L)^2]^{1/2}} v_i e^{j(\omega t - \phi_1)}$$

From Eq. 2.16:

$$v_o(t) = L \frac{di}{dt} = \frac{R - j\omega L}{R^2 + (\omega L)^2} (j\omega L) v_i e^{j\omega t} = \frac{(\omega L)^2 + j\omega LR}{R^2 + (\omega L)^2} v_i e^{j\omega t}$$

$$v_o = \sqrt{(Re)^2 + (Im)^2}$$
$$= \frac{[(\omega L)^4 + (\omega LR)^2]^{1/2}}{R^2 + (\omega L)^2} v_i = \frac{\omega L}{[R^2 + (\omega L)^2]^{1/2}} v_i$$

$$\phi_2 = \tan^{-1} \frac{Im}{Re} = \frac{\omega LR}{(\omega L)^2} = \frac{R}{\omega L}$$

$$v_o(t) = v_o e^{j(\omega t - \phi_2)} = \frac{\omega L}{[R^2 + (\omega L)^2]^{1/2}} v_i e^{j(\omega t - \phi_2)}$$

Exercise 2.26

For the equivalent circuit for Fig. E2.26:

$$\frac{1}{R_{e1}} = \frac{1}{R_1} + \frac{1}{R_2} + \frac{1}{R_3}$$

$$R_{e1} = \frac{R_1 R_2 R_3}{R_2 R_3 + R_1 R_3 + R_1 R_2}$$

$$C_e = C_1 + C_2$$

$$\frac{1}{R_{e2}} = \frac{1}{R_4} + \frac{1}{R_5}$$

$$R_{e2} = \frac{R_4 R_5}{R_4 + R_5}$$

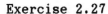

From Eq. 2.49:
$$Z = \left[\left(R_{e1} + R_{e2}\right)^2 + \left(\omega L - \frac{1}{\omega C_e}\right)^2\right]^{1/2}$$

From Eq. 2.50:
$$\phi = \tan^{-1} \frac{\omega L - 1/\omega C_e}{R_{e1} + R_{e2}}$$

Exercise 2.27

For the circuit shown in Fig. 2.14:

From Eq. 2.49:

$$Z(\omega) = \sqrt{R^2 + (\omega L - 1/\omega C)^2}$$

Case	L (μH)	C (μF)	R (kΩ)
(a)	0.01	0.50	10,000
(b)	0.05	0.20	1,000
(c)	0.10	0.10	500
(d)	0.20	0.20	200
(e)	0.50	0.50	100

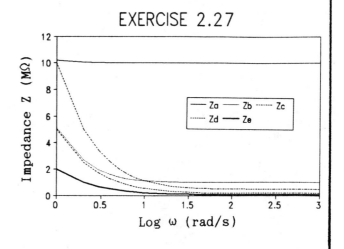

Exercise 2.28

For the circuit shown in Fig. 2.13:

From Eqs. 2.54 and 2.55:

$$|H(\omega)| = \frac{1}{\sqrt{1 + (\omega RC)^2}}$$

$$\phi = -\tan^{-1} \omega RC$$

Case	C (μF)	R (MΩ)
(a)	0.50	10
(b)	0.20	1
(c)	0.10	0.50
(d)	0.20	0.20
(e)	0.50	0.10

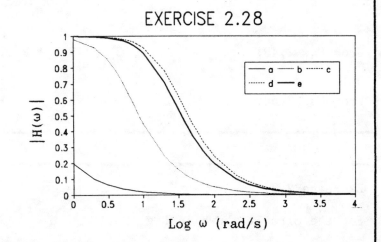

Exercise 2.29

From Eqs. 2.56 and 2.57:

$$N_{dB} = 10 \log (p/p_r) \qquad N_{dB} = 20 \log (v/v_r)$$

Case	p/p_r	N_{dB}	Case	v/v_r	N_{dB}
(a)	1000	30	(a)	15	23.52
(b)	2.0	3.01	(b)	0.001	-60
(c)	0.003	-25.2	(c)	3000	69.5

Exercise 2.30

From Eq. 2.54:

$$|H(\omega)| = \frac{1}{\sqrt{1 + (\omega RC)^2}}$$

From Eq. 2.58:

$$N_{dB} = -10 \log [1 + (\omega RC)^2]$$

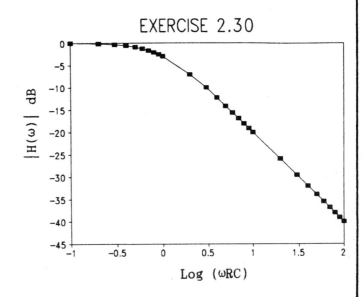

Exercise 2.31

From Eq. 2.55:

$$\phi = -\tan^{-1} \omega RC$$

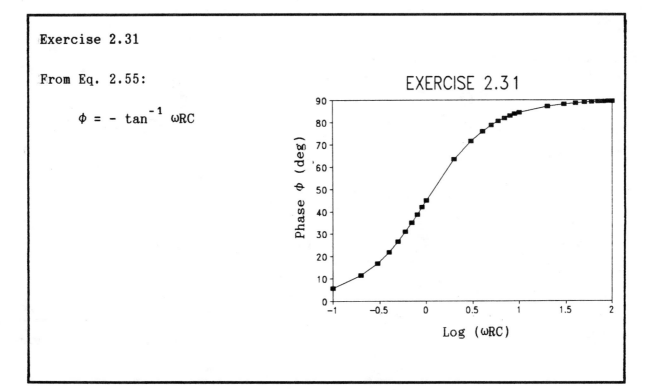

Exercise 2.32

$$|H(\omega)| = \frac{1}{\sqrt{1 + (R/\omega L)^2}}$$

From Eq. 2.58:

$$N_{dB} = -10 \log [1 + (R/\omega L)^2]$$

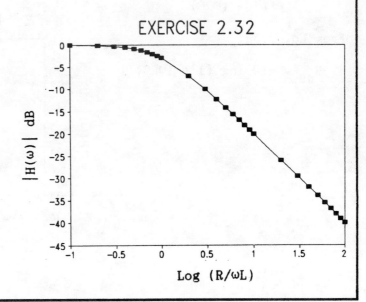

Exercise 2.33

From Eq. 2.55:

$$\phi = - \tan^{-1} (R/\omega L)$$

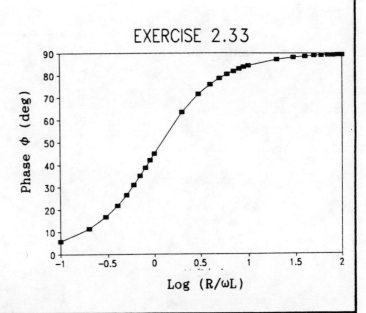

Exercise 3.1

1. Input impedance
2. Sensitivity
3. Range
4. Zero drift
5. Frequency response

Exercise 3.2

From Eq. (3.1):
$$p = \frac{v^2}{Z_m}$$

v (V)	Power loss p (W)				
	Input impedance Z_m (Ω)				
	(a) 10^3	(b) 10^4	(c) 10^5	(d) 10^6	(e) 10^7
0	0	0	0	0	0
5	0.025	0.0025	0.0003	0.00003	0.000003
10	0.100	0.0100	0.0010	0.00010	0.000010
20	0.400	0.0400	0.0040	0.00040	0.000040
30	0.900	0.0900	0.0090	0.00090	0.000090
50	2.500	0.2500	0.0250	0.00250	0.000250
75	5.625	0.5625	0.0563	0.00563	0.000563
100	10.000	1.0000	0.1000	0.01000	0.001000

The data can be plotted in a useful form by using a log-log plot. Thus,

From Eq. 3-1: $p = v^2/Z_m$ $\quad \log p = 2 \log v - \log Z_m$

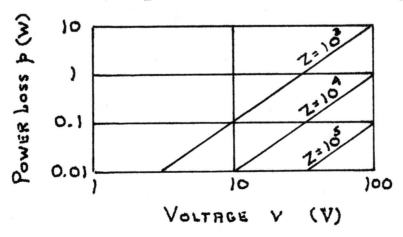

Exercise 3.3

From Eq. 3.2b with $R_m = 10^6 \, \Omega$ and $C = 100(10^{-12})$ F:

$$R_m C = 10^6 (100)(10^{-12}) = 100(10^{-6}) \text{ s} = 100 \, \mu s$$

$$|Z_m| = \frac{R_m}{\sqrt{1 + (\omega R_m C)^2}}$$

Therefore,

$$\frac{|Z|}{R_m} = \frac{1}{\sqrt{1 + (\omega R_m C)^2}}$$

$$\omega_C = \frac{1}{R_m C} = \frac{1}{100(10^{-6})} = 10{,}000 \text{ rad/s}$$

$$f_C = \frac{\omega_C}{2\pi} = \frac{10{,}000}{2\pi} = 1{,}592 \text{ Hz}$$

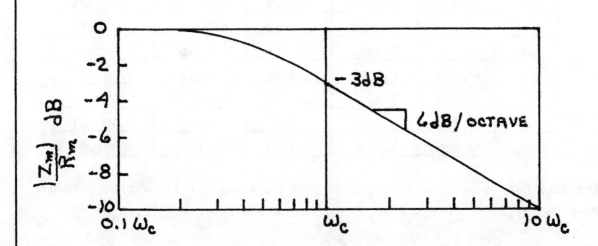

Exercise 3.4

From Eq. 3.2a:

$$Z_m = \frac{R_m}{1 + j\omega R_m C}$$

$$v_m = Z_m i = \left(\frac{Z_m}{R_s + Z_m}\right) v_i$$

$$\text{Error} = \frac{v_i - v_m}{v_i} = 1 - \frac{Z_m}{R_s + Z_m} = \frac{R_s}{R_s + Z_m} = \frac{(R_s/R_m)(1 + j\omega R_m C)}{1 + (R_s/R_m)(1 + j\omega R_m C)}$$

With $R_s = 10\ \Omega$, $R_m = 10^6\ \Omega$, and $C = 100$ pF:

$$(R_s/R_m) = 10/10^6 = 10(10^{-6})$$

$$R_m C = 10^6(100)(10^{-12}) = 100(10^{-6})\ \text{s}$$

$$\text{Error} = \frac{(R_s/R_m)(1 + j\omega R_m C)}{1 + (R_s/R_m)(1 + j\omega R_m C)} = \frac{10(10^{-6})[1 + j(100)(10^{-6})\omega]}{1 + 10(10^{-6})[1 + j(100)(10^{-6})\omega]}$$

$$\cong 10(10^{-6})[1 + j100(10^{-6})\omega]$$

(a) For $\omega = 10^2$:

$$\mathcal{E} \cong 10(10^{-6})[1 + j100(10^{-6})(10^2)]$$

$$\cong 0.00001(1 + 0.01j) \cong 0.00001 = 0.001\%$$

(b) For $\omega = 10^4$:

$$\mathcal{E} \cong 10(10^{-6})[1 + j100(10^{-6})(10^4)]$$

$$\cong 0.00001(1 + j) \cong 0.000014 = 0.0014\% \text{ (with } 45° \text{ phase shift)}$$

(c) For $\omega = 10^6$:

$$\mathcal{E} \cong 10(10^{-6})[1 + j100(10^{-6})(10^6)]$$

$$\cong 0.00001(1 + 100j) \cong 0.001 = 0.1\% \text{ (with } 90° \text{ phase shift)}$$

Exercise 3.5

From Eq. (3.4):
$$\mathcal{E} = \frac{R_s/R_m}{1 + (R_s/R_m)}$$

Solving for R_s/R_m:

(a) For $\mathcal{E} = 0.5\%$:
$$\frac{R_s}{R_m} = \frac{\mathcal{E}}{1 - \mathcal{E}} = \frac{0.005}{1 - 0.005} = 0.00503$$

(b) For $\mathcal{E} = 1.0\%$:
$$\frac{R_s}{R_m} = \frac{\mathcal{E}}{1 - \mathcal{E}} = \frac{0.01}{1 - 0.01} = 0.0101$$

(c) For $\mathcal{E} = 2.0\%$:
$$\frac{R_s}{R_m} = \frac{\mathcal{E}}{1 - \mathcal{E}} = \frac{0.02}{1 - 0.02} = 0.0204$$

(d) For $\mathcal{E} = 5.0\%$:
$$\frac{R_s}{R_m} = \frac{\mathcal{E}}{1 - \mathcal{E}} = \frac{0.05}{1 - 0.05} = 0.0526$$

Exercise 3.6

From Eq. (3.6):
$$S_R = \frac{1}{S} \qquad v_1 = d\, S_R$$

(a) For $d = 2.5$ div and $S = 0.2$ div/V:
$$S_R = \frac{1}{S} = \frac{1}{0.2} = 5 \text{ V/Div}$$
$$v_1 = d\, S_R = 2.5(5) = 12.5 \text{ V}$$

(b) For $d = 6.1$ div and $S = 0.5$ div/V:
$$S_R = \frac{1}{S} = \frac{1}{0.5} = 2 \text{ V/Div}$$
$$v_1 = d\, S_R = 6.1(2) = 12.2 \text{ V}$$

(c) For $d = 3.7$ div and $S = 2.0$ div/V:
$$S_R = \frac{1}{S} = \frac{1}{2.0} = 0.5 \text{ V/Div}$$
$$v_1 = d\, S_R = 3.7(0.5) = 1.85 \text{ V}$$

(d) For $d = 7.6$ div and $S = 0.1$ div/V:
$$S_R = \frac{1}{S} = \frac{1}{0.1} = 10 \text{ V/Div}$$
$$v_1 = d\, S_R = 7.6(10) = 76 \text{ V}$$

ENGINEERING MEASUREMENTS by J. W. DALLY, W. F. RILEY, AND K. G. McCONNELL

Exercise 3.7

From Eq. (3.7): $\quad v^* = \dfrac{d^*}{S}$

(a) For $S = 0.2$: $\quad v^* = \dfrac{d^*}{S} = \dfrac{8}{0.2} = 40$ V

(b) For $S = 0.5$: $\quad v^* = \dfrac{d^*}{S} = \dfrac{8}{0.5} = 16$ V

(c) For $S = 2.0$: $\quad v^* = \dfrac{d^*}{S} = \dfrac{8}{2.0} = 4$ V

(d) For $S = 0.1$: $\quad v^* = \dfrac{d^*}{S} = \dfrac{8}{0.1} = 80$ V

Since $v^*S = d^*$ (a constant), range v^* can be increased only if sensitivity S is decreased.

Exercise 3.8

Range of an instrument system represents the maximum value that can be measured. Sensitivity $S = d/v_c$, where d is the pen displacement that can be accurately measured. The relation between range and sensitivity $v^* = d^*/S$ shows the tradeoff between these characteristics. When the sensitivity S is high, the range v^* will be low and vice versa.

Exercise 3.9

$$15 \text{ days} = 15(24) = 360 \text{ hours}$$

$$\text{Maximum drift} = 0.05(360) = 18.0 \text{ mV}$$

$$\%\mathcal{E} = \dfrac{18.0}{12}(100) = 150\%$$

Exercise 3.10

From Eq. (3.10):

$$N_{dB} = 20 \log_{10}(A_o/A_i) = -1.5$$

$$\log_{10}(A_o/A_i) = \frac{-1.5}{20} = -0.075$$

$$A_o/A_i = 10^{-0.075} = 0.841$$

$$\%\varepsilon = \frac{1 - 0.841}{1}(100) = 15.9\%$$

Exercise 3.11

From Eq. (3.10): $N_{dB} = 20 \log_{10}(C_o/C_i)$

For $f = 10$ Hz: $N_{dB} = 20 \log_{10}(1.01) = 0.0864$

For $f = 20$ Hz: $N_{dB} = 20 \log_{10}(1.03) = 0.2567$

For $f = 40$ Hz: $N_{dB} = 20 \log_{10}(1.05) = 0.4238$

For $f = 60$ Hz: $N_{dB} = 20 \log_{10}(1.00) = 0.0000$

For $f = 80$ Hz: $N_{dB} = 20 \log_{10}(0.93) = -0.6303$

For $f = 100$ Hz: $N_{dB} = 20 \log_{10}(0.80) = -1.9382$

Exercise 3.12

From Eq. (3.10):

(a) $N_{dB} = 20 \log_{10}(A_o/A_i) = -5.0$

$\log_{10}(A_o/A_i) = \frac{-5.0}{20} = -0.250$

$A_o/A_i = 10^{-0.250} = 0.562$

(b) $N_{dB} = 20 \log_{10}(A_o/A_i) = 2.5$

$\log_{10}(A_o/A_i) = \frac{+2.5}{20} = +0.125$

$A_o/A_i = 10^{+0.125} = 1.334$

(c) $N_{dB} = 20 \log_{10}(A_o/A_i) = -0.25$

$\log_{10}(A_o/A_i) = \frac{-0.25}{20} = -0.0125$

$A_o/A_i = 10^{-0.0125} = 0.972$

(d) $N_{dB} = 20 \log_{10}(A_o/A_i) = +0.70$

$\log_{10}(A_o/A_i) = \frac{+0.70}{20} = +0.035$

$A_o/A_i = 10^{+0.035} = 1.084$

Exercise 3.13

$$\%\mathcal{E} = \frac{1 - (A_o/A_i)}{1}(100) \quad \text{or} \quad (A_o/A_i) = 1 - \frac{\%\mathcal{E}}{100}$$

(a) For $\mathcal{E} = \pm 5\%$: $(A_o/A_i) = 1 - \frac{\%\mathcal{E}}{100} = 1 - \frac{\pm 5}{100} = 0.95$ or 1.05

$N_{dB} = 20 \, \text{Log}_{10} \, (0.95) = -0.446$

$N_{dB} = 20 \, \text{Log}_{10} \, (1.05) = +0.424$

$-0.446 < N_{dB} < +0.424$

(b) For $\mathcal{E} = \pm 2\%$: $(A_o/A_i) = 1 - \frac{\%\mathcal{E}}{100} = 1 - \frac{\pm 2}{100} = 0.98$ or 1.02

$N_{dB} = 20 \, \text{Log}_{10} \, (0.98) = -0.1766$

$N_{dB} = 20 \, \text{Log}_{10} \, (1.02) = +0.1720$

$-0.1766 < N_{dB} < +0.1720$

(c) For $\mathcal{E} = \pm 1\%$: $(A_o/A_i) = 1 - \frac{\%\mathcal{E}}{100} = 1 - \frac{\pm 1}{100} = 0.99$ or 1.01

$N_{dB} = 20 \, \text{Log}_{10} \, (0.99) = -0.0873$

$N_{dB} = 20 \, \text{Log}_{10} \, (1.01) = +0.0864$

$-0.0873 < N_{dB} < +0.0864$

(d) For $\mathcal{E} = \pm 0.5\%$: $(A_o/A_i) = 1 - \frac{\%\mathcal{E}}{100} = 1 - \frac{\pm 0.5}{100} = 0.995$ or 1.005

$N_{dB} = 20 \, \text{Log}_{10} \, (0.995) = -0.0435$

$N_{dB} = 20 \, \text{Log}_{10} \, (1.005) = +0.0433$

$-0.0435 < N_{dB} < +0.0433$

Exercise 3.14

From Eq. (3.13): $\theta = Si$

$$S = \frac{\theta}{i} = \frac{45}{25} = 1.80 \text{ deg}/\mu A = 0.0314 \text{ rad}/\mu A$$

Exercise 3.15

From Eq. (3.15):
$$R_{sh} = \frac{i_m^*}{i_i^* - i_m^*} R_m$$

(a) For $i_i^* = 100$ mA: $\quad R_{sh} = \frac{0.010}{0.100 - 0.010}(40) = 4.444\ \Omega$

(b) For $i_i^* = 0.5$ A: $\quad R_{sh} = \frac{0.010}{0.5 - 0.010}(40) = 0.816\ \Omega$

(c) For $i_i^* = 2$ A: $\quad R_{sh} = \frac{0.010}{2 - 0.010}(40) = 0.201\ \Omega$

(d) For $i_i^* = 20$ A: $\quad R_{sh} = \frac{0.010}{20 - 0.010}(40) = 0.020\ \Omega$

Exercise 3.16

From Eq. (3.16):
$$R_{sr} = \frac{v^*}{i_m^*} - R_m$$

$$= \frac{v^*}{0.000050} - 40 = 20{,}000\,v^* - 40$$

For $v^* = 10$ mV: $\quad R_{sr} = 20{,}000(0.010) - 40 = 160\ \Omega$

Similarly:

v^* (V)	R_{sr} (Ω)
0.01	160
0.05	960
0.10	1,960
0.50	9,960
1.00	19,960
5.00	99,960
10.00	199,960
50.00	999,960
100.00	1,999,960

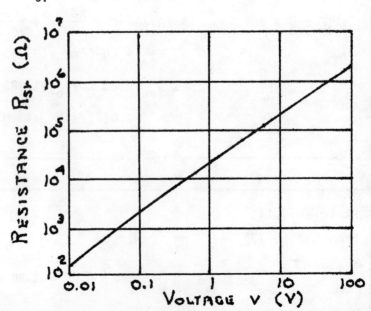

Exercise 3.17

From Eq. (3.17): $\quad \mathcal{E} = \dfrac{R_s/(R_m + R_{sr})}{1 + R_s/(R_m + R_{sr})}$

For $R_s = 10\ \Omega$, $R_m = 40\ \Omega$, and $R_{sr} = 160\ \Omega$:

$$\mathcal{E} = \dfrac{10/(40 + 160)}{1 + 10/(40 + 160)} = 0.04762 = 4.762\%$$

Similarly:

v^* (V)	R_{sr} (Ω)	$\mathcal{E}$ (%)
0.01	160	4.762
0.05	960	0.990
0.10	1,960	0.498
0.50	9,960	0.100
1.00	19,960	0.050
5.00	99,960	0.010
10.00	199,960	0.005
50.00	999,960	0.001
100.00	1,999,960	0.0005

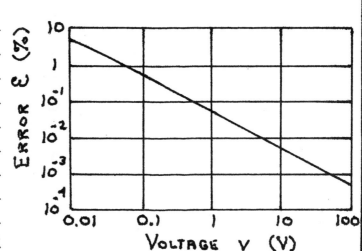

Exercise 3.18

1. Amplifier permits an increase in R_{sr} by a factor equal to the gain.
2. Source resistance R_s can be increased by a factor equal to the gain while maintaining the sensitivity.
3. Source resistance can be increased without increasing measurement error.

Exercise 3.19

Advantages:

1. Accurate (±0.005%) measurement of voltage can be made without associated load error.
2. Method is very sensitive yet inexpensive.

Disadvantages:

1. Method of measurement is very time consuming.

ENGINEERING MEASUREMENTS by J. W. DALLY, W. F. RILEY, AND K. G. McCONNELL

Exercise 3.20

From Eq. (3.21): $\quad v_x = \frac{x}{\ell} v_r$

With $\ell = 10$ in. and $v_r = 2$ mV:

(a) For $x = 4.5$ in. $\quad v_x = \frac{4.5}{10}(2) = 0.90$ mV

(b) For $x = 7.2$ in. $\quad v_x = \frac{7.2}{10}(2) = 1.44$ mV

(c) For $x = 2.1$ in. $\quad v_x = \frac{2.1}{10}(2) = 0.42$ mV

(d) For $x = 8.6$ in. $\quad v_x = \frac{8.6}{10}(2) = 1.72$ mV

Exercise 3.21

Different signals can be supplied to the recorder with a multiplexer (or switch) in sequence. Different signals are identified by a different color or character.

Exercise 3.22

Deadband: The small zone about the balance point where friction prevents exact zeroing of the error signal.

Slewing Speed: Maximum velocity of the pen when both servos are moving the pen at maximum speed.

Exercise 3.23

Assume that a strain extensiometer is attached to a tensile specimen and that the specimen is being extended in a testing machine equipped with a load cell. The signal from the load cell is recorded on the y-axis and the signal from the extensiometer is recorded on the x-axis. The gain on the y-axis is adjusted to give a convenient scale for the stress $\sigma = P/A_o$ where A_o is the initial cross-sectional area of the tensile specimen.

Similarly the gain on the x-axis is adjusted to give a convenient scale to the strain $\varepsilon = \Delta \ell / \ell_o$ where ℓ_o is the gage length of the extensiometer (usually 2 in.). The modulus of elasticity, yield strength, and ultimate tensile strength are read from the $\sigma - \varepsilon$ diagram.

ENGINEERING MEASUREMENTS by J. W. DALLY, W. F. RILEY, AND K. G. McCONNELL

Exercise 3.24

A galvanometer is used to drive the pen or the mirror in pen type and light-writing type oscillographs. A synchronous motor is used to drive the paper at a constant velocity so that position along the paper corresponds to time.

With the pen type oscillograph, current flow through the galvanometer causes coil rotation which in turn drives a short arm carrying the pen. Ink from the pen traces a record on graph paper along the y-axis indicating the current (or voltage) imposed on the galvanometer coil. Since the paper under the pen is moving at a constant velocity, the ink trace represents a y-t record of the event.

The light-writing oscillograph is identical except that a mirror is driven by the galvanometer instead of a pen. A light beam is focused on the mirror, and as it rotates, the light beam traces the y-t record on light-sensitive paper.

The pen type oscillograph has a low frequency response (about 100 Hz) because of the inertia of the lever arm and the heavier mechanism required to drive the pen assembly. It is also less sensitive because of the relatively short lever arm.

The light writing oscillograph is capable of higher frequencies (up to 25 kHz) because the mirror is small and the support assembly can be light with low inertia. This type oscillograph is also more sensitive because the lever arm, which is a folded light beam, can be made very long to provide a large amplification of the coil motion.

Exercise 3.25

(a) For f_{max} = 80 Hz: Use an M200-120

(b) For f_{max} = 300 Hz: Use an M400-350

(c) For f_{max} = 800 Hz: Use an M1650

(d) For f_{max} = 9000 Hz: Use an M13000

Exercise 3.26

(a) For the M200-120 Galvanometer: R_x = 120 Ω   S = 1.61 cm/mV

(b) For the M400-350 Galvanometer: R_x = 350 Ω   S = 0.264 cm/mV

(c) For the M1650 Galvanometer: R_x = 100 Ω   S = 10.2 cm/V

(d) For the M13000 Galvanometer: R_x = 100 Ω   S = 0.435 cm/V

Exercise 3.27

From Eq. (3.32):
$$\ddot{\theta} + 2d\omega_n \dot{\theta} + \omega_n^2 \theta = \omega_n^2 SI_s$$

Complementary Solution: Assume $\theta_c = \theta_0 e^{\lambda t}$

$$\lambda^2 \theta_0 e^{\lambda t} + 2d\omega_n \lambda e^{\lambda t} + \omega_n^2 \theta_0 e^{\lambda t} = 0$$

$$\lambda^2 + 2d\omega_n + \omega_n^2 = 0$$

Which has the solution:

$$\lambda = -\omega_n d \pm \omega_n \sqrt{d^2 - 1}$$

$$\theta_c = A e^{-(d+\sqrt{d^2-1})\omega_n t} + B e^{-(d-\sqrt{d^2-1})\omega_n t}$$

Steady State Solution:
$$\theta_s = SI_s$$
$$\theta = \theta_c + \theta_s$$

Initial Conditions: At $t = 0$, $\theta = \dot{\theta} = 0$

$$A + B + \theta_s = 0$$

$$\lambda_1 A e^{\lambda_1 t} + \lambda_2 B e^{\lambda_2 t} = 0$$

Solving Simultaneously:

$$A = -\frac{\lambda_2 \theta_s}{\lambda_2 - \lambda_1} \qquad B = \frac{\lambda_1 \theta_s}{\lambda_2 - \lambda_1}$$

Where: For $d > 1$

$$\lambda_1 = \left(-d + \sqrt{d^2 - 1}\right)\omega_n$$

$$\lambda_2 = \left(-d - \sqrt{d^2 - 1}\right)\omega_n$$

$$\lambda_2 - \lambda_1 = -2\sqrt{d^2 - 1}$$

$$\frac{\theta}{\theta_s} = 1 + \frac{\lambda_1}{\lambda_2 - \lambda_1} e^{\lambda_2 t} - \frac{\lambda_2}{\lambda_2 - \lambda_1} e^{\lambda_1 t}$$

ENGINEERING MEASUREMENTS by J. W. DALLY, W. F. RILEY, AND K. G. McCONNELL

Exercise 3.28

See Exercise 3.27 for solution of Eq. (3.32):

For $d = 1$:
$$\lambda_1 = \lambda_2 = -d\omega_n = -\omega_n$$

$$\theta = A e^{\lambda t} + Bt\, e^{\lambda t} + \theta_s$$

Initial Conditions: At $t = 0$, $\theta = \dot{\theta} = 0$

$$A + \theta_s = 0 \quad \text{therefore} \quad A = -\theta_s$$

$$\lambda A + B = 0 \quad \text{therefore} \quad B = -\lambda A = \omega_n \theta_s$$

$$\frac{\theta}{\theta_s} = 1 - e^{-\omega_n t} - \omega_n t\, e^{-\omega_n t} = 1 - (1 + \omega_n t)\, e^{-\omega_n t}$$

Exercise 3.29

See Exercise 3.27 for solution of Eq. (3.32):

For $d < 1$: $\quad \lambda_1 = -d\omega_n + i\omega_d \quad\quad \lambda_2 = -d\omega_n - i\omega_d$

Where: $\quad \omega_d = \omega_n \sqrt{1 - d^2} \quad\quad i = \sqrt{-1}$

$$\theta_c = A_1 e^{\lambda_1 t} + B_1 e^{\lambda_2 t} = e^{-d\omega_n t}\left[A_1 e^{i\omega_d t} + B_1 e^{-i\omega_d t}\right]$$

$$= e^{-d\omega_n t}\left[A \cos \omega_d t + B \sin \omega_d t\right]$$

$$\theta = \theta_c + \theta_s$$

Initial Conditions: At $t = 0$, $\quad \theta = \dot{\theta} = 0$

$$A + \theta_s = 0 \quad \text{therefore} \quad A = -\theta_s$$

$$-d\omega_n A + B\omega_d = 0 \quad\quad B = \frac{d}{\sqrt{1-d^2}} A = -\frac{d\,\theta_s}{\sqrt{1-d^2}}$$

Therefore:

$$\frac{\theta}{\theta_s} = 1 - e^{-d\omega_n t}\left[\cos \omega_n \sqrt{1-d^2}\, t + \frac{d}{\sqrt{1-d^2}} \sin \omega_n \sqrt{1-d^2}\, t\right]$$

Exercise 3.30

From Eq. (3.36):
$$\frac{\theta}{\theta_s} = 1 - (1 + \omega_n t) e^{-\omega_n t}$$

For $\mathcal{E} = \pm 2\%$:
$$\frac{\theta}{\theta_s} = 0.98 \text{ or } 1.02$$

$\mathcal{E} = 1.02$ is not possible for a critically damped galvanometer.

Recall that:
$$T_n = \frac{2\pi}{\omega_n} \quad \text{or} \quad \omega_n = \frac{2\pi}{T_n}$$

Therefore:
$$\frac{\theta}{\theta_s} = 1 - \left(1 + 2\pi \frac{t}{T_n}\right) e^{-2\pi(t/T_n)} = 0.98$$

Which has the solution:
$$\frac{t}{T_n} = 0.929 \cong 0.93$$

Exercise 3.31

From Eq. (3.37):

$$Z = 1 - \frac{e^{-d\omega_n t}}{\sqrt{1 - d^2}} \cos(\omega_d t - \phi)$$

where $\tan \phi = \dfrac{d}{\sqrt{1 - d^2}}$

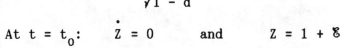

At $t = t_0$: $\dot{Z} = 0$ and $Z = 1 + \mathcal{E}$

$$\dot{Z} = d\omega_n \cos(\omega_d t_0 - \phi) + \omega_d \sin(\omega_d t_0 - \phi) = 0$$

Which yields:
$$\tan(\omega_d t_0 - \phi) = -\frac{d}{\omega_d/\omega_n} = -\frac{d}{\sqrt{1 - d^2}} = -\tan \phi$$

Therefore, $\omega_d t_0 = \pi$ and with $Z = 1 + \mathcal{E}$

$$\mathcal{E} = -\frac{e^{-d\omega_n t_0}}{\sqrt{1 - d^2}} \cos(\pi - \phi) = -\frac{e^{-(d/\sqrt{1-d^2})\pi}}{\sqrt{1 - d^2}}\left(-\sqrt{1 - d^2}\right) = e^{-(d/\sqrt{1-d^2})\pi}$$

Thus,
$$\mathcal{E} e^{(d/\sqrt{1-d^2})\pi} = 1 \quad \text{or} \quad d^2 = \frac{\ln^2 \mathcal{E}}{\pi^2 + \ln^2 \mathcal{E}}$$

Exercise 3.32

From Eq. 3.40:
$$\frac{1}{\omega_n^2}\ddot{\theta} + \frac{2d}{\omega_N}\dot{\theta} + \theta = \theta_a e^{j\omega t}$$

where: $\quad \theta_a = si_a$

Assume: $\quad \theta = Ae^{j\omega t}$

Differentiating yields:
$$\dot{\theta} = Aj\omega e^{j\omega t}$$
$$\ddot{\theta} = -A\omega^2 e^{j\omega t}$$

Substituting into Eq. 3.40:
$$A[-(\omega/\omega_n)^2 + j\, 2d(\omega/\omega_n) + 1] = \theta_a = si_a = \theta_s$$

Let $r = \omega/\omega_n$,
$$A = \frac{\theta_s}{(1-r^2) + j2dr} = \frac{(1-r^2) - j\,2dr}{(1-r^2)^2 + (2dr)^2}\theta_s$$

Then:
$$\frac{\theta}{\theta_s} = \left[\frac{1-r^2}{B} - j\,\frac{2dr}{B}\right]e^{j\omega t}$$

where
$$B = \frac{1}{(1-r^2)^2 + 2dr)^2}$$

$$A_0 = |H(\omega)| = \sqrt{(Re)^2 + (Im)^2} = B\left[(1-r^2)^2 + (2dr)^2\right]^{1/2} = \frac{1}{\sqrt{B}}$$

$$\phi = \tan^{-1}\frac{Im}{Re} = \tan^{-1}\frac{-2dr}{1-r^2}$$

$$\frac{\theta}{\theta_s} = H(\omega) = A_0 e^{-j\phi}$$

Exercise 3.33

From Eq. 3.43 with $r = \omega/\omega_n$:

$$a^2(1 - r^2) + 4a^2d^2r^2 = 1 \qquad (1)$$

For maximum a: $\qquad \dfrac{da}{dr} = 0$

Which yields: $\qquad r^2 = 1 - 2d^2 \qquad (2)$

Substituting (2) into (1) yields:

$$d^2 = \tfrac{1}{2} - \tfrac{1}{2}\sqrt{1 - 1/a^2}$$

For $\mathscr{E} = 10\%$ $\quad a = 1.10$

$$d^2 = \tfrac{1}{2} - \tfrac{1}{2}\sqrt{1 - 1/(1.1)^2} = 0.2917$$

Therefore: $\qquad d = 0.540$

Value of r at the lower error limit is obtained from Eq. (1). Thus,

$$(1 - r^2)^2 + 4d^2r^2 = \dfrac{1}{a^2}$$

With $\mathscr{E} = 10\%$, $\quad a = 0.90 \quad$ and $\quad d = 0.540$

$$(1 - r^2)^2 + 4(0.54)^2 r^2 = \left(\dfrac{1}{0.90}\right)^2$$

$$r^4 - 0.8336r^2 - 0.2346 = 0$$

$$r^2 = 1.0558 \text{ and } -0.2222$$

$$r = 1.028$$

Similarly for other error values:

% $\mathscr{E}$	d	(ω/ω_n)max
± 10	0.540	1.028
± 5	0.589	0.870
± 2	0.634	0.692
± 1	0.655	0.585

Exercise 3.34

$$i = 10 \sin 400t + 2 \sin 800t + \sin 1200t$$

(a) For $d = 0.55$ and $\omega_n = 1100$ rad/sec:

$$r_1 = \frac{400}{1100} = 0.3636, \quad r_2 = \frac{800}{1100} = 0.7273, \quad r_3 = \frac{1200}{1100} = 1.0909$$

From Eqs. 3.43 and 3.44:

$a_1 = 1.047(10) = 10.47$ \hspace{2cm} $\phi_1 = 24.74° = 0.432$ rad.

$a_2 = 1.077(2) = 2.154$ \hspace{2cm} $\phi_2 = 59.51° = 1.039$ rad.

$a_3 = 0.823(1) = 0.823$ \hspace{2cm} $\phi_3 = 99.00° = 1.728$ rad.

$$i_o = 10.47 \sin(400t - 0.432) + 2.154 \sin(800t - 1.039)$$
$$+ 0.823 \sin(1200t - 1.728)$$

(b) For $d = 0.60$ and $\omega_n = 1200$ rad/sec:

$$r_1 = \frac{400}{1200} = 0.3333, \quad r_2 = \frac{800}{1200} = 0.6667, \quad r_3 = \frac{1200}{1200} = 1.000$$

From Eqs. 3.43 and 3.44:

$a_1 = 1.026(10) = 10.26$ \hspace{2cm} $\phi_1 = 24.23° = 0.423$ rad.

$a_2 = 1.027(2) = 2.054$ \hspace{2cm} $\phi_2 = 55.23° = 0.964$ rad.

$a_3 = 0.833(1) = 0.833$ \hspace{2cm} $\phi_3 = 90.00° = 1.571$ rad.

$$i_o = 10.26 \sin(400t - 0.423) + 2.054 \sin(800t - 0.964)$$
$$+ 0.833 \sin(1200t - 1.571)$$

(c) For $d = 0.65$ and $\omega_n = 1300$ rad/sec:

$$r_1 = \frac{400}{1300} = 0.3077, \quad r_2 = \frac{800}{1300} = 0.6154, \quad r_3 = \frac{1200}{1300} = 0.9231$$

From Eqs. 3.43 and 3.44:

$a_1 = 1.010(10) = 10.10$ \hspace{2cm} $\phi_1 = 23.84° = 0.416$ rad.

$a_2 = 0.987(2) = 1.974$ \hspace{2cm} $\phi_2 = 52.17° = 0.911$ rad.

$a_3 = 0.827(1) = 0.827$ \hspace{2cm} $\phi_3 = 82.97° = 1.446$ rad.

$$i_o = 10.10 \sin(400t - 0.416) + 1.974 \sin(800t - 0.911)$$
$$+ 0.827 \sin(1200t - 1.446)$$

Exercise 3.35

Internal: Triggers on y-signal presented to the screen; good for low level signals.

Line: Useful when measuring periodic waveforms that exhibit a fundamental frequency of 60 Hz.

External: Independent trigger from an external source. Used for measurement of transient events.

Exercise 3.36

A cathode-ray tube (CRT) is a vacuum tube used to display voltage-time traces. A focused beam of electrons (negatively charged particles) creates the display. The electrons are generated by heating a cathode that emits a cloud of electrons. A series of hollow anodes (positively charged plates) collect the electrons, form them into a beam, and accelerate them toward the face of the tube. The beam strikes the face of the tube and is absorbed by the thin layer of fluorescent material. The electron bombardment of the fluorescent material generates photons which produce a bright point of light that locates the electron beam. The electron beam is used to display voltage-time records on the face of the CRT. This is accomplished by inserting pairs of horizontal and vertical plates in the CRT tube adjacent to the electron beam. Voltages imposed on the horizontal plates either attract or repel the electron beam producing y-deflections on the face of the CRT. Similarly, voltages on the vertical plates produce x-deflections of the electron beam.
In this configuration, the CRT provides the display for an x-y recorder. Its advantage is its extremely high frequency response because of the nearly insignificant mass of the electrons.

Exercise 3.37

A general purpose oscilloscope is a complex instrument with one or more y-amplifiers, a z-amplifier, and a time base. The amplifiers can be single ended or differential, can be ac or dc, and can have a choice of 10 or more gains. The time base usually has a wide choice of sweep speeds. Trigger options are available for repetitive or single sweeps. This instrument is designed for many different applications and is normally used by an operator familiar with amplifiers, CRTs, and triggers.
The hospital oscilloscope is different. It triggers automatically and the controls are limited and simple. The transducers and oscilloscopes are matched to eliminate the need for a variety of controls. This instrument is designed for an operator not familiar with electronic measurements.

Exercise 3.38

The hospital oscilloscope is designed to monitor human body functions which are (1) relatively low frequency and (2) nearly the same for two different individuals. (Imagine your heart rate increasing by an order of magnitude).

The measurements are made primarily to monitor body functions; therefore, simplicity and ease of reading are essential (scales are usually removed from the face of the CRT).

The general purpose oscilloscope is designed for use in a wide range of applications. Precise data can be obtained with many different transducers.

No attempt is made to match the oscilloscope with any transducer. The versatility of the instrument leads to complexity but operators are expected to understand its components and to make detailed and accurate measurements.

Exercise 3.39

Bandwidth: Frequency range over which signals are recorded with less than a 3-dB loss compared to midband performance. The higher the frequency response the shorter the rise time that the system can respond to.

$$f_{bw} t_r = 1.70$$

Exercise 3.40

From Eq. (3.49):
$$t_r = \frac{1.70}{f_{bw}}$$

or
$$\text{Log } t_r = \text{Log } 1.70 - \text{Log } f_{bw}$$

ENGINEERING MEASUREMENTS by J. W. DALLY, W. F. RILEY, AND K. G. McCONNELL

Exercise 3.41

A portable oscilloscope is small and light weight so that it may be carried into the field. Also, it is battery operated and power limited. To achieve a small, light weight, and low-power instrument, differences are required in the designs of the portable oscilloscopes. The CRT is smaller and the amplifiers are simpler with fewer selections of gain or sweep rate. Miniature switches and control knobs are used. Rechargable battery packs are used for power supplies.

Exercise 3.42

AM Recording Advantages:

1. Simple, low cost.
2. Suitable for audio signals.

AM Recording Disadvantages:

1. Cannot record signals with frequencies between dc (0) and 50 Hz.
2. Tape imperfections often produce reductions in signal level for short periods of time.
3. Can only record and store data. No display.

Exercise 3.43

FM Recording Advantages:

1. Low frequency and dc signals can be recorded.
2. Tape imperfections do not seriously affect the output.

FM Recording Disadvantages:

1. Much more expensive than AM equipment.
2. Can only record and store data. Data must be displayed with other equipment such as oscillographs or oscilloscopes.

Exercise 3.44

Digital Recording Advantages:

1. Easier to use than FM recording since tape speed is not critical.
2. Amplifiers used are simple and inexpensive.
3. Output data is in a form suitable for processing with a computer.

Digital Recording Disadvantages:

1. Input signal must be digitized prior to recording.
2. Sensitive to tape dropout; therefore, high quality tape must be used.

Exercise 3.45

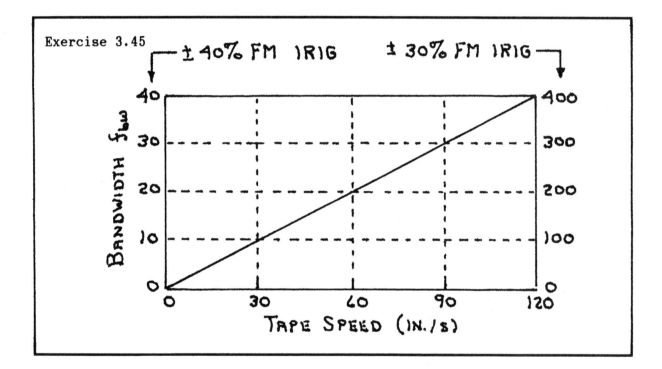

Exercise 3.46

The 400 word summary should include paragraphs discussing the following items:
1. The major importance of frequency response which divides recording instruments into three or four different categories.
2. The general characteristics of recording instruments which include input impedance, sensitivity, range, and frequency response.
3. Steady state instruments.
4. Instruments for slowly varying signals.
5. Instruments for rapidly varying signals.

Exercise 3.47

$$|Z_m| = \frac{R_m}{\sqrt{1 + (\omega R_m C)^2}} \qquad (3.2b)$$

$$\mathcal{E} = \frac{R_s/(R_m + R_{sr})}{1 + R_s/(R_m + R_s)} \qquad (3.17)$$

$$\frac{\theta}{\theta_s} = 1 - \frac{e^{-d\omega_n t}}{\sqrt{1 - d^2}} \cos(\omega_d t - \phi) \qquad (3.37)$$

$$\frac{\theta}{\theta_s} = H(\omega) = \frac{e^{-j\phi}}{\sqrt{[1 - (\omega/\omega_n)^2]^2 + 4d^2(\omega/\omega_n)^2}} \qquad (3.45)$$

$$f_{bw} t_r = 1.70 \qquad (3.49)$$

Exercise 3.48

Eq. (3.2b):   Input impedance of a recording instrument.

Eq. (3.17):   Loading error for a voltmeter.

Eq. (3.37):   Response of a 2nd-order system to a step input.

Eq. (3.45):   Frequency response of a 2nd-order system (periodic).

Eq. (3.49):   Bandwidth-rise time relationship for an amplifier.

Exercise 4.1

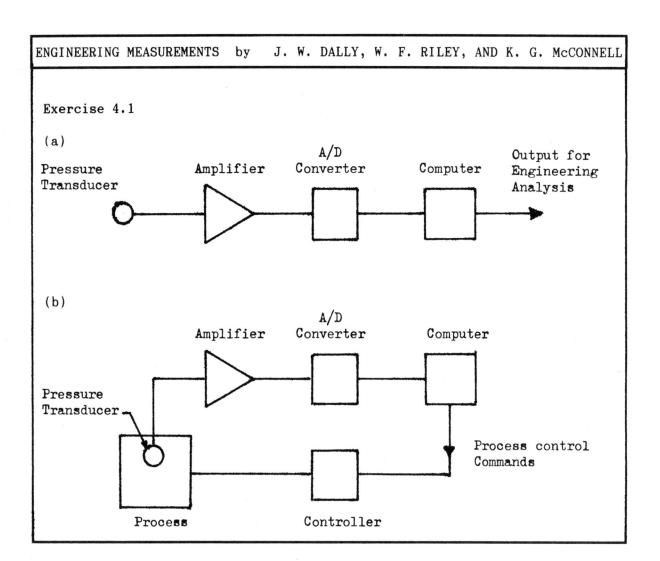

Exercise 4.2

From Eq. 4.1: $\qquad C = (2^n - 1)$

N	C	N	C	N	C	N	C
1	1	9	511	17	131,071	25	33,554,431
2	3	10	1023	18	262,143	26	67,128,864
3	7	11	2047	19	524,287	27	134,217,727
4	15	12	4095	20	1,048,575	28	268,435,455
5	31	13	8191	21	2,097,151	29	536,870,911
6	63	14	16,383	22	4,194,303	30	1,073,741,823
7	127	15	32,767	23	8,388,607	31	2,147,483,647
8	255	16	65,535	24	16,777,215	32	4,294,967,295

Exercise 4.3

From Eq. 4.2a:
$$R = \frac{2^0}{2^n - 1} = \frac{1}{2^n - 1} = \frac{1}{C}$$

N	C	R
1	1	1
4	15	$6.667(10^{-2})$
8	255	$3.922(10^{-3})$
12	4095	$2.442(10^{-4})$
16	65,535	$1.526(10^{-5})$
20	1,048,575	$9.537(10^{-7})$
24	16,777,215	$5.960(10^{-8})$
28	268,435,455	$3.725(10^{-9})$
32	4,294,967,295	$2.328(10^{-10})$

Exercise 4.4

$$\mathcal{E}_R (\%) = 100\, R$$

N	C	R	$\mathcal{E}_R$ (%)
1	1	1	100
4	15	$6.667(10^{-2})$	6.667
8	255	$3.922(10^{-3})$	$3.922(10^{-1})$
12	4095	$2.442(10^{-4})$	$2.442(10^{-2})$
16	65,535	$1.526(10^{-5})$	$1.526(10^{-3})$
20	1,048,575	$9.537(10^{-7})$	$9.537(10^{-5})$
24	16,777,215	$5.960(10^{-8})$	$5.960(10^{-6})$
28	268,435,455	$3.725(10^{-9})$	$3.725(10^{-7})$
32	4,294,967,295	$2.328(10^{-10})$	$2.328(10^{-8})$

Exercise 4.5

Table 4.2 indicates that an instrument with 8-bit circuits will provide a 0.392 % resolution error at best.

Exercise 4.6

An analog-to-digital converter is an electronic device that takes an analog signal as an input and converts it to a digital code. Binary numbers are usually employed as the digital code since the numbers required 0 and 1 can be represented by an open or closed switch. There are three different conversion methods in common use today: —

   (1) successive approximation method.
   (2) dual slope integration method.
   (3) parallel or flash conversion method.

With the successive approximation method, the unknown analog voltage is compared to known voltages generated in a logic controlled sequence by a DAC. When the comparison is within ±1/2 LSB, the approximation is terminated and the input to the DAC is the output for the analog-to-digital converter.

With the dual slope integration method, the unknown voltage is added to a reference voltage and the sum is integrated for a fixed time. The reference voltage is then switched on the integrator and the output voltage (from the integrator) decreases linearly with time. The time of the zero crossing of the output voltage is related to the unknown voltage. A clock is used to measure and to provide a digital code for the unknown voltage.

The flash converter employs ($2^n - 1$) voltage comparators arranged in parallel. Each comparator is connected to the unknown voltage v_u but the reference voltage is applied through a binary resistance ladder such that the reference voltage v_R^* between adjacent comparators varies by 1 LSB. The digital code representing v_u is developed by logic circuits after the comparator where $v_R^* > v_u$ has been determined.

Exercise 4.7

A digital-to-analog converter is an electronic device that takes a digital signal and converts it to an equivalent analog voltage. This conversion is accomplished by bringing the digital signal into the DAC on a parallel bus. The signal is used to switch (or not switch) logic gates that connect the reference voltage to an n-bit summing amplifier. The resistances in each of the n lines feeding the summing amplifier are arranged in a binary sequence 1, 2, 4, 8, etc. The output from the summing amplifier gives an analog voltage that is scaled relative to the reference voltage by the digital input signal (code).

ENGINEERING MEASUREMENTS by J. W. DALLY, W. F. RILEY, AND K. G. McCONNELL

Exercise 4.8

Resolution $R = 1/C$ where $C = 2^n - 1$ is the count. For large n, $R \approx 2^{-n}$ and resolution error $\mathcal{E}_R \approx 2^{-n}(100)$.

Quantization error $\pm(1/2)$LSB is more important than resolution error. It can be shown that the standard deviation of a number of measurements is $1/2\sqrt{3}$ which indicates that the effective resolution of an ADC is much better than the resolution error described in the previous paragraph.

Exercise 4.9

From Eq. 4.2b:
$$\sigma = \frac{2^{-(n+1)}}{\sqrt{3}}$$

$$\log \sigma = \log \frac{1}{\sqrt{3}} - (n+1) \log 2 = K - (n+1)) K_1$$

$$K = -0.239 \qquad K_1 = 0.301$$

$$\log \sigma = -0.239 - 0.301 (n+1)$$

$$= -0.540 - 0.301 n$$

N	log σ	σ
6	-2.346	$4.508(10^{-3})$
8	-2.948	$1.127(10^{-3})$
10	-3.550	$2.818(10^{-4})$
12	-4.152	$7.047(10^{-5})$
14	-4.754	$1.762(10^{-5})$
16	-5.356	$4.406(10^{-6})$

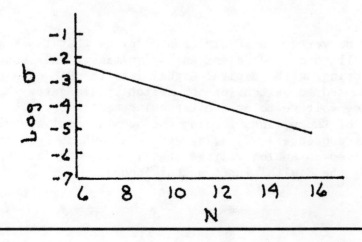

Exercise 4.10

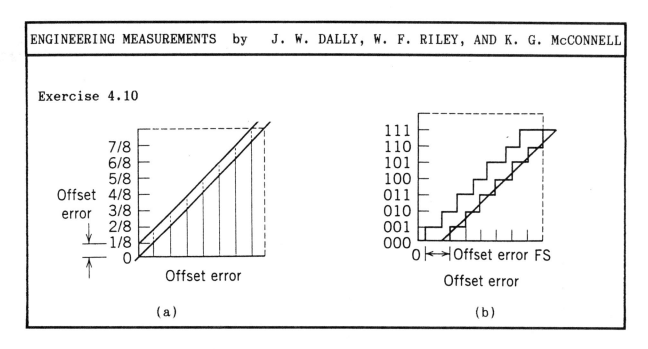

Exercise 4.11

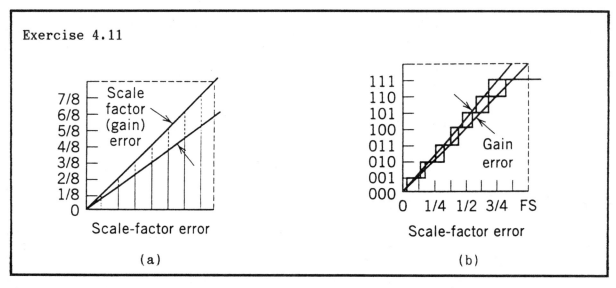

Exercise 4.12

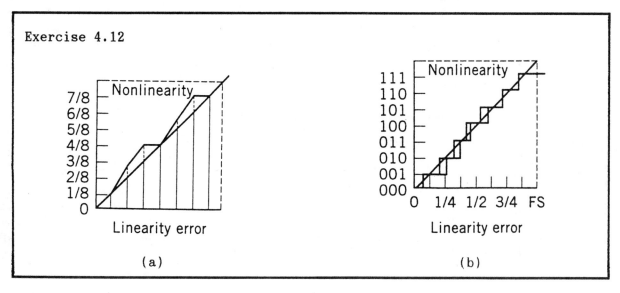

Exercise 4.13

From Eq. 4.5:
$$v_o = -R_f \sum_{n=1}^{4} (R_f/R_i)$$

For the 4-bit DAC shown in Fig. 4.5:

$$v_o = 10\left[\frac{5}{10} + \frac{5}{20} + \frac{5}{40} + \frac{5}{80}\right] = 10\left[\frac{1}{2} + \frac{1}{4} + \frac{1}{8} + \frac{1}{16}\right]$$

(a) $\quad 1101 \rightarrow 10\left[\frac{1}{2} + \frac{1}{4} + 0 + \frac{1}{16}\right] = 8.125$ volts

(b) $\quad 1010 \rightarrow 10\left[\frac{1}{2} + 0 + \frac{1}{8} + 0\right] = 6.250$ volts

(c) $\quad 0110 \rightarrow 10\left[0 + \frac{1}{4} + \frac{1}{8} + 0\right] = 3.750$ volts

(d) $\quad 0101 \rightarrow 10\left[0 + \frac{1}{4} + 0 + \frac{1}{16}\right] = 3.125$

(e) $\quad 1001 \rightarrow 10\left[\frac{1}{2} + 0 + 0 + \frac{1}{16}\right] = 5.625$

(f) $\quad 1110 \rightarrow 10\left[\frac{1}{2} + \frac{1}{4} + \frac{1}{8} + 0\right] = 8.75$ volts

---

Exercise 4.14

A register is a small memory element that will accept the output from a device like an ADC and store it temporarily. A typical register stores at least one bit but rarely more than a few words. In an ADC, the digital code in the register is stored until it is ready to be used in activating the display.

Exercise 4.15

Analog-to-digital converters sample, convert, store, and display in a time-controlled sequence of operations. Each operation requires a specified time. A clock is incorporated in the system to keep time and issue signals which control the conversion process. The strobe is part of the controlling logic that activates the process of transferring the digital signal from the register to the display.

ENGINEERING MEASUREMENTS by J. W. DALLY, W. F. RILEY, AND K. G. McCONNELL

Exercise 4.16

Successive approximation: medium speed, medium accuracy, medium cost.

Dual slope integration: low speed, high accuracy, low cost.

Flash or parallel: high speed, low accuracy, high cost.

Exercise 4.17

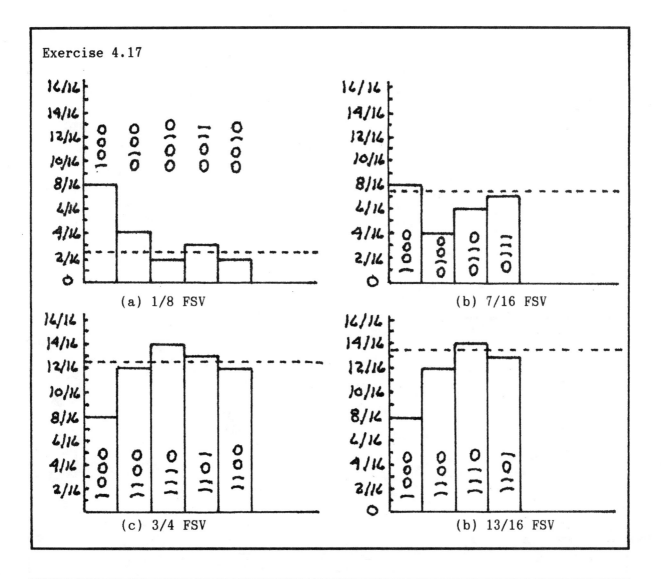

Exercise 4.18

The A/D converter incorporated 4 bits ($2^4 = 16$). All conversions were an exact fraction of 16.

Exercise 4.19

A sample of the unknown voltage is collected in a very short period of time by a sample-and-hold amplifier. This voltage is held (maintained at a constant value) for the time required for conversion. A new sample is then taken and the process is repeated.

Exercise 4.20

In Fig. 4.8, the input voltage (unknown) is sampled in a very short period of time and maintained at that value. This signal is transmitted to a voltage comparator where it is held constant during the conversion process. Control logic inputs a mid-scale digital code 1000 into a DAC and the output from the DAC goes to the other input of the voltage comparator. The voltage comparison indicates if the digital code (the initial approximation) was high or low. The logic control circuit adjusts the digital code to give a better approximation and the process is repeated for each of the n bits in the A/D converter. At the end of the process, the final digital code is obtained and provided as output from the register.

Exercise 4.21

From Eq. 4.10:
$$v_i = \frac{(v_u + v_R)t^*}{2RC}$$

From Eq. 4.11:
$$\Delta v_i = - v_R \frac{\Delta t}{RC}$$

At the zero crossing:

$$v_o = v_i + \Delta v_i = v_i - v_R \frac{\Delta t}{RC} = 0$$

Thus,

$$\frac{v_u + v_R}{2RC} t^* = v_R \frac{\Delta t}{RC}$$

From which,

$$\frac{\Delta t}{t^*} = \frac{1}{2}\left(\frac{v_u}{v_R} + 1\right)$$

Exercise 4.22

The summing amplifier is used to combine reference and unknown voltages v_R and v_u. This combined signal is integrated in an integrating amplifier for a specific time t^* and the charging ramp is produced. At $t = t^*$, the summing amplifier is disconnected from the integrator and the reference voltage is connected directly to the input of the integrator. The voltage out of the integrator decreases at a controlled rate and the discharge ramp is produced. The time Δt associated with the zero crossing of the discharge ramp is determined with a voltage comparator and measured by the number of clock pulses needed to give Δt. The number of clock pulses is related to the digital conversion of v_u.

Exercise 4.23

From Eq. 4.13:
$$v_u = \frac{v_u^*}{2} \sin 2\pi f t$$

$$\frac{dv_u}{dt} = \frac{2\pi f v_u^*}{2} \cos 2\pi f t$$

$$\text{slew rate} = \left(\frac{dv_u}{dt}\right)_{max} = \pi f v_u^*$$

Recall that $\Delta v < (\text{LSB} \times \text{FSV})$ during the conversion time T.

$$\Delta V_{max} = \pi f v_u^* T < (\text{LSB} \times \text{FSV}) = 2^{-n} \text{FSV}$$

Since
$$v_u^* = \text{FSV}$$

$$f < \frac{2^{-n}}{\pi T}$$

Exercise 4.24

For a 12-bit unipolar converter (n = 12) capable of 20 readings per second,

$$T = \frac{1}{20} = 0.05 \text{ s}$$

From Eq. 4.16:
$$f < \frac{2^{-n}}{\pi T} < \frac{2^{-12}}{\pi(0.05)} = 0.001554 \text{ Hz}$$

ENGINEERING MEASUREMENTS by J. W. DALLY, W. F. RILEY, AND K. G. McCONNELL

Exercise 4.25

For a 12-bit bipolar converter (n = 13) capable of 20 readings per second,

$$T = \frac{1}{20} = 0.05 \text{ s}$$

From Eq. 4.16:
$$f < \frac{2^{-n}}{\pi T} < \frac{2^{-13}}{\pi(0.05)} = 0.000777 \text{ Hz}$$

Exercise 4.26

With $f = 0.001554$ Hz (see Exercise 4.24) and $v_u^* = $ FSV:

From Eq. 4.14:
$$\left(\frac{dv_u}{dt}\right)_{max} = \pi f v_u^* = \pi(0.001554)(\text{FSV}) = 0.004882(\text{FSV})$$

(a) $v_u^* = 1$ V: $\left(\frac{dv_u}{dt}\right)_{max} = 0.004882(\text{FSV})$
$$= 0.004882(1) = 4.882(10^{-3}) \text{ V/s} = 4.882 \text{ mV/s}$$

(b) $v_u^* = 2$ V: $\left(\frac{dv_u}{dt}\right)_{max} = 0.004882(\text{FSV})$
$$= 0.004882(2) = 9.764(10^{-3}) \text{ V/s} = 9.764 \text{ mV/s}$$

(c) $v_u^* = 5$ V: $\left(\frac{dv_u}{dt}\right)_{max} = 0.004882(\text{FSV})$
$$= 0.004882(5) = 24.41(10^{-3}) \text{ V/s} = 24.41 \text{ mV/s}$$

(d) $v_u^* = 10$ V: $\left(\frac{dv_u}{dt}\right)_{max} = 0.004882(\text{FSV})$
$$= 0.004882(10) = 48.82(10^{-3}) \text{ V/s} = 48.82 \text{ mV/s}$$

Exercise 4.27

Parallel A/D converters contains a total of $2^n - 1$ voltage comparators arranged in parallel. The unknown analog voltage v_u is connected to one of the inputs in each comparator. The other input in each comparator is connected to the reference voltage v_R through a binary ladder resistance. The comparator where $v_u > v_R$ changes to $v_u < v_R$ can be detected. The logic gates are arranged to convert this comparator number to a digital code that becomes the output.

Exercise 4.28

Flash type A/D converters are employed in digital oscilloscopes and wave form recorders where high frequency analog signals are converted into digital signals, displayed as discrete voltage time traces, and stored. The sampling rate and the bandwidth are the same, say 10 MHz. This sampling rate implies a sampling time of 0.1 μs.

Exercise 4.29

When several signals are being processed by a single system, each signal must be identified and tracked for the entire event. The address bus is used to control the flow of data (like traffic lights controlling traffic) and to place the data in the proper registers. Data buses are the wires (usually parallel wires) over which the data is transmitted (like the streets over which traffic flows).

Exercise 4.30

In Fig. 4.12, m analog inputs are available as inputs to be monitored by a single system. A program in the microprocessor (input by an operator) controls the multiplex through the address bus. Analog inputs (voltages) are switched into the differential amplifier at specified times consistent with the conversion speed of the ADC. The differential amplifier increases the voltages to levels consistent with the full scale range of the ADC. The sample and hold amplifier takes each voltage in turn for a very short period and holds it constant while the ADC converts it to digital format. The digital signal is output into a register in the interface identified through the address bus. The process is repeated for each of the inputs. The data is held in the interface in temporary but identified registers. Instructions programmed in the microprocessor and transmitted by the address bus move the data over the data bus to permanent storage locations in RAM or elsewhere.

Exercise 4.31

Parallel bus structures are the fastest way to move data but noise and attenuation limit transmission to short distances (tens of feet).

A serial bus is used when transmission distances become large. These buses are low loss cables protected from noise (coaxial cable or fiber optic). Also, amplifying stations at periodic intervals refresh the signals.

ENGINEERING MEASUREMENTS   by   J. W. DALLY, W. F. RILEY, AND K. G. McCONNELL

Exercise 4.32

The RS-232 is a standard bus structure that employs a serial protocol where a single transmitter sends one bit (0 or 1) of information at a time to a single receiver. This method of transmission is used when data-transfer rates are low or when data must be transferred over long distances. The serial protocol can be converted by a modem, placed on a standard telephone line, and converted back to serial data by a second modem at the receiving end of the line. The RS-232 bus structure is popular because most personal computers have at least one RS-232 port and many digital instruments are available with an RS-232 bus structure.

Exercise 4.33

The IEEE-488 bus is a parallel protocol that transmits eight bits of information simultaneously over a standard cable. The bus contains eight data lines, three handshake lines, five interface-management lines, and eight ground and shield wires. The bus carries information between a system controller and one or more digital instruments. Unlike the RS-232 bus, which is limited to a single instrument, the IEEE-488 bus can connect up to 15 compatible instruments simultaneously.

Exercise 4.34

RS-232:

Advantages:	Port available on most digital instruments.
	Common and popular.
Disadvantages:	Low baud rate.
	Single receiver only.

IEEE 488:

Advantages:	Ease of use.
	Fewer transmission errors.
	High baud rate.
Disadvantages:	Port often not available.
	Length limit on multiple devices.

Exercise 4.35

The data is input directly to the board. The address and bus structures employed reside in the PC except for those supplied on the board.

Exercise 4.36

Main power at ±5 V and ±15 V, fans, bus structures, cabinets, RAM, displays, etc. are all contained in the PC system. No costs are incurred for any of these items.

Exercise 4.37

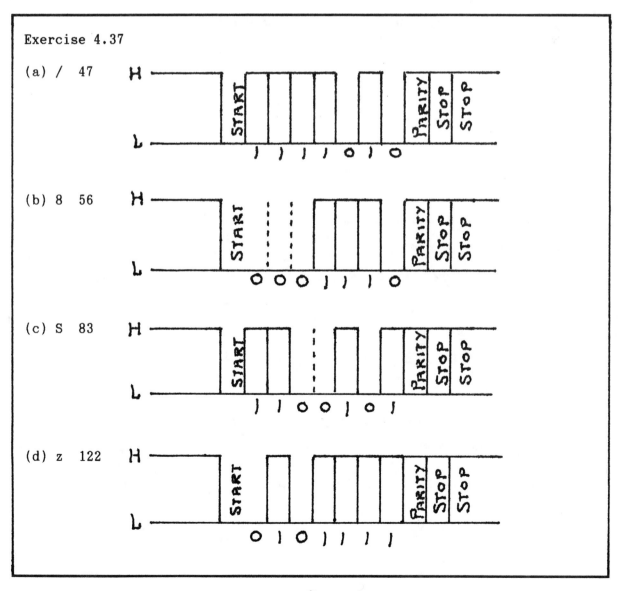

Exercise 4.38

Baud rate, the speed of transmission of a digital signal, is measured in k bits/s. It depends on the digital devices at both ends of the line, the capabilities of the transmission line, and the distance transmitted.

Exercise 4.39

The maximum count for a 5 1/2 - digit DVM is:

(a)   99,999   With 0 % overranging

(b)   199,999   With 100 % overranging

(c)   299,999   With 200 % overranging

Exercise 4.40

For a 5 1/2 digit DVM with 100 % overranging maximum count is 199,999.

For a specified accuracy of ±0.002 % ± 2 counts:

(a) $1.80000 \pm (0.000036 + 0.00002) \Rightarrow \begin{matrix} 1.80006 \\ 1.79994 \end{matrix}$

(b) $2.50000 \pm (0.000050 + 0.0002) \Rightarrow \begin{matrix} 2.5003 \\ 2.4997 \end{matrix}$

(c) $9.99996 \pm (0.000199 + 0.0002) \Rightarrow \begin{matrix} 10.0004 \\ 9.9995 \end{matrix}$

Exercise 4.41

See Exercise 4.40 for maximum and minimum values of v_o (readings).

(a) $\mathcal{E} = \dfrac{v_o - v_i}{v_i}(100) = \dfrac{1.80006 - 1.80000}{1.80000}(100) = +0.0033\%$

$\mathcal{E} = \dfrac{v_o - v_i}{v_i}(100) = \dfrac{1.79994 - 1.80000}{1.80000}(100) = -0.0033\%$

(b) $\mathcal{E} = \dfrac{v_o - v_i}{v_i}(100) = \dfrac{2.5003 - 2.50000}{2.50000}(100) = +0.0120\%$

$\mathcal{E} = \dfrac{v_o - v_i}{v_i}(100) = \dfrac{2.4997 - 2.50000}{2.50000}(100) = -0.0120\%$

(c) $\mathcal{E} = \dfrac{v_o - v_i}{v_i}(100) = \dfrac{10.0004 - 9.99996}{9.99996}(100) = +0.0044\%$

$\mathcal{E} = \dfrac{v_o - v_i}{v_i}(100) = \dfrac{9.9995 - 9.99996}{9.99996}(100) = -0.0046\%$

Exercise 4.42

System multimeters are more complex than bench-type multimeters. Both instruments convert an analog signal to a digital reading and display the reading. However, a system multimeter contains a microprocessor, local memory, and a bus to facilitate interfacing with other components of an automated data-processing system.

Exercise 4.43

The specification should include detailed requirements for:

1. The number of digits and overranging and conversion rate.
2. The maximum count and resolution.
3. The sensitivity and the ranges required.
4. The accuracy required.
5. The calibration requirement.
6. The stability requirements.
7. Overload protection requirements.
8. Microprocessor requirements.
9. System bus outputs.

ENGINEERING MEASUREMENTS by J. W. DALLY, W. F. RILEY, AND K. G. McCONNELL

Exercise 4.44

A system multimeter is a component in a data-logging system. The data-logging system also contains a multiplexer, an interface, an active and long term memory, and a printer. The data-logging system also incorporates a microprocessor and an address bus for controlling the acquisition, conversion, storage, and display of the data.

Exercise 4.45

ROM (read only memory) is often used to store permanent instructions in the microprocessor control device in a data-logging system.

RAM (random access memory) is used to store temporary instructions and to store data in registers or in interfaces until it can be downloaded into more permanent memory.

Exercise 4.46

A scanner, also known as a multiplexer, is a programmable switch. It is used as the interface between large numbers of transducers and the data-logging or data-acquisition system. Upon command, the scanner switches from one signal to another at a rate consistent with the capabilities of the data- logging or data-acquisition system.

Exercise 4.47

Data-logging and data-acquisition systems are similar in that they accept a large number of analog input signals, convert them to digital signals, and begin the processing or storage of these signals. There are two principal differences. The data-acquisition system is much faster with sample rates from 20,000 to 250,000 conversions per second. Second, the on board microcomputer, memory capacity, and speed (clock rate) is superior to that available with a data-logging system. Of course the costs of the data- acquisition systems are significantly higher.

ENGINEERING MEASUREMENTS by J. W. DALLY, W. F. RILEY, AND K. G. McCONNELL

Exercise 4.48

1. Controller.
2. Signal conditioner.
3. Multiplex-amplifier.
4. Analog-to-digital converter.
5. Storage (memory unit).
6. Readout devices.

Exercise 4.49

The controller is a microprocessor which serves as the interface between the operator and the data-acquisition system. The operator programs the controller by selecting sample rate, sequence of channels to be monitored, signal levels to start and stop processing, and signal levels to activate alarms. The controller also directs the flow of data, its storage, and its retention.

Exercise 4.50

Section 1. Signal conditioning, multiplexing, and amplification.
2. Analog-to-digital conversion.
3. Microcomputer for data processing and data flow.
4. Bus interface to facilitate transfer to host PC.

Exercise 4.51

Digital and analog oscilloscopes are similar except that the digital oscilloscope can:

1. Store the converted digital signal.
2. Manipulate that signal before displaying it.
3. Process the signal on-board.

The digital oscilloscope is superior in every respect to the analog oscilloscope except for very high speed transients with rise times that are small compared to the conversion times. Of course digital oscilloscopes are more costly than analog oscilloscopes.

Exercise 4.52

1. Permanent inexpensive storage of data.
2. Store many events.
3. Average data from several different events.
4. Provide ability to compare events.

Exercise 4.53

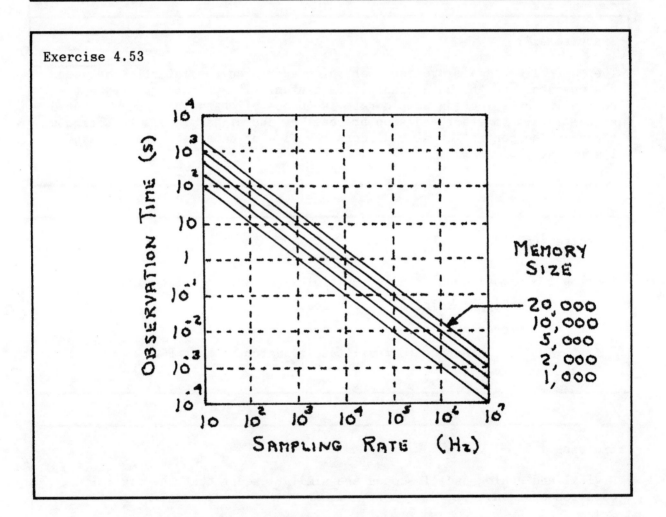

Exercise 4.54

Repetitive signals are easier to measure since sampling can be repeated on the second and subsequent waveforms to give apparent sampling rates that are an order of magnitude higher than the real sampling rates. A delayed sequential sampling technique, illustrated in Fig. 4.24, is employed to increase the number of samples that define the repetitive waveform.

ENGINEERING MEASUREMENTS by J. W. DALLY, W. F. RILEY, AND K. G. McCONNELL

Exercise 4.55

Data is converted and stored continuously in an auxiliary buffer. This data is discarded if a trigger signal does not occur during a given sweep (filling of the buffer). If a trigger occurs during the filling period, the entire buffer is retained and the data stored before and after the trigger is available in memory.

Exercise 4.56

Modern digital oscilloscopes are usually equipped with a microprocessor that provides several on-board, signal-analysis features. These features often include the following:

1. Pulse characterization - rise time, fall time, base-line and top-line width, overshoot, period, frequency, rms, standard deviation, and duty cycle.
2. Frequency analysis - power, phase, and magnitude spectrum.
3. Spectrum analysis - 100 to 50,000 point fast Fourier transforms.
4. Math package - add, subtract, multiply, integrate, and differentiate.
5. Smoothing - 1, 3, 5, 7, or 9 point.
6. Counter - average frequency and event crossings.
7. Display control - x zoom, x position, y gain, y offset.
8. Plotting display.
9. Mass storage to floppy disk, hard disk, or nonvolatile bubble memory.

Exercise 4.57

Wave form recorders do not have a CRT display; however, they usually have superior resolution and larger memories than digital oscilloscopes with comparable sampling rates.

Exercise 4.58

From Eqs. 4.20 and 4.22:

$$f_2 = 2mf_c \pm f_1 \qquad f_c = f_s/2$$

$$f_2 = 2m(f_s/2) \pm f_1$$

Thus,

$$\pm f_1 = f_2 - mf_s \quad \text{provided } f_2 > f_s/2$$

Exercise 4.59

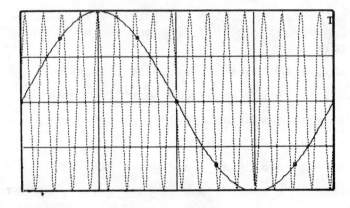

(a) —— alias signal, ···· original signal

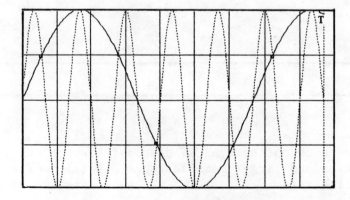

(b) —— alias signal, ···· original signal

Exercise 4.59 (Continued)

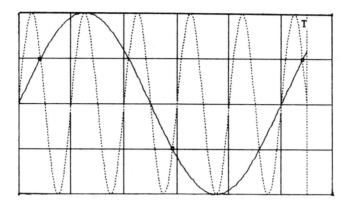

(c) —— alias signal, ···· original signal

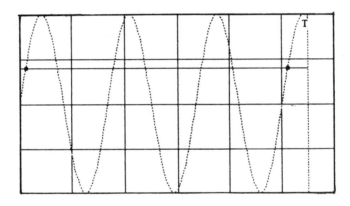

(d) —— alias signal, ···· original signal

Since f_2 is an integer multiple of f_s, the alias signal is a constant.

Exercise 4.60

Anti-aliasing filters cut out frequencies in the input signal that exceed the Nyquist frequency.

Exercise 4.61

The advantage is that false signals generated by aliasing in A to D conversion cannot occur. The disadvantage is that high frequency components filtered out of the signal may be real in which case the data recorded may be in serious error.

ENGINEERING MEASUREMENTS by J. W. DALLY, W. F. RILEY, AND K. G. McCONNELL

Exercise 5.1

Sensors are electrical components that provide a voltage output when subjected to a change in configuration or environment. For example, piezoelectric and piezoresistive sensors respond to a change in pressure. Transducers usually incorporate a mechanical element that controls the change in configuration or environment on the sensor. In an accelerometer, a mass controls the force (pressure) acting on the piezo-electric sensor. A pressure transducer with a differential transformer to sense the displacement of the center point of the diaphragm, is an example of a transducer incorporating a displacement sensor. A strain gage is an example of a sensor, often employed in load transducers, that is also a strain transducer.

Exercise 5.2

Number of turns = n

$$n = \frac{L}{d} = \frac{100}{0.1} = 1000 \text{ turns}$$

$$\text{Resolution} = \frac{L}{n} = \frac{100}{1000} = 0.10 \text{ mm}$$

Exercise 5.3

Recall:
$$p = \frac{v^2}{R}$$

Therefore:
$$v_{max} = \sqrt{pR} = \sqrt{2(2000)} = 63.2 \text{ V}$$

$$\Delta v = \frac{63.2}{100/0.10} = 0.0632 \text{ V} = 63.2 \text{ mV}$$

Exercise 5.4

$$N = 100(20) = 2000 \text{ Divisions.}$$

$$\Delta R = \frac{20,000}{2000} = 10 \text{ } \Omega = \text{Dial Resolution}$$

$$\Delta R = 20,000(0.0005) = 10 \text{ } \Omega = \text{Potentiometer Resolution}$$

Therefore: $\Delta R = 10 \text{ } \Omega$ (Larger value if the two are not equal)

ENGINEERING MEASUREMENTS by J. W. DALLY, W. F. RILEY, AND K. G. McCONNELL

Exercise 5.5

1. Inertia of the parts. 2. Wiper bounce.

Exercise 5.6

1. Near infinite resolution (Limits due to nonlinearities)
2. Lower noise
3. Longer life (less wear)

Exercise 5.7

$L = 100$ m $L_c > 100$ m
P = Pulley PP = Pulley with potentiometer
E = Elevator r = Pulley radius
S = Spring to maintain cable tension to eliminate slip.

$$\frac{100}{20} = 5 \text{ m/turn} \qquad r = \frac{5}{2\pi} = 0.796 \text{ m}$$

For a resolution of 10 mm:

$$\frac{100(10^3)}{20(10)} = 500 \text{ div/turn}$$

A high quality thin-film potentiometer with a resolution of 0.002 % should be used.

Exercise 5.8

Characteristic	Potentiometer	LVDT
Range	Large	Small
Accuracy	Good	Fair
Resolution	Good	Good
Freq. Response	Poor	Fair
Reliability	Excellent	Good
Complexity	Simple	Moderate
Cost	Low	Moderate

Exercise 5.9

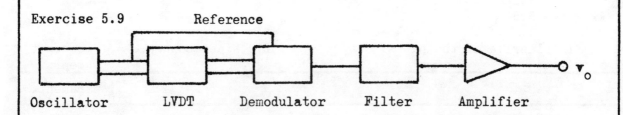

Oscillator   LVDT   Demodulator   Filter   Amplifier

Oscillator: - Drives primary coil with constant magnitude ac voltage and provides reference signal for demodulator circuit.

LVDT: - Has a primary and two secondary coils that are connected by a movable magnetic link.

Demodulator: - Converts the high frequency output signal into a low frequency phase sensitive signal that is proportional to the position of the magnetic core.

Filter: - Low pass filter to remove high frequency carrier (oscillator) signals.

Amplifier: - Increases system sensitivity and couples the LVDT to a recorder.

Exercise 5.10

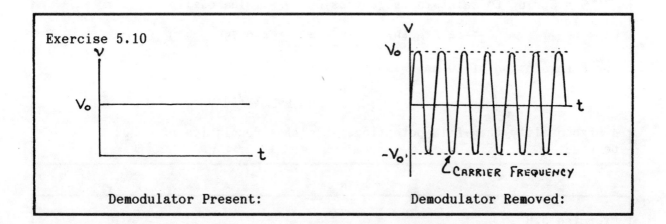

Demodulator Present: Demodulator Removed:

Exercise 5.11

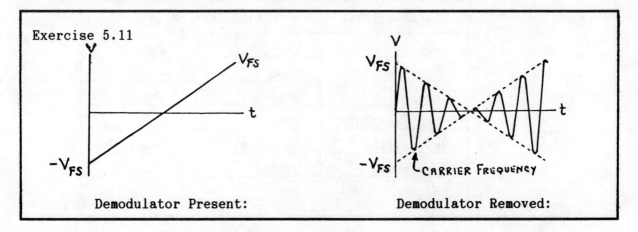

Demodulator Present: Demodulator Removed:

Exercise 5.12

LVDT: All electronic components required for operation of this device must be provided since they are not an integral part of the instrument.

DCDT: All electronic components required for operation of this instrument have been installed within the case of the instrument. Only a dc voltage must be supplied when a DCDT is used to make measurements.

Exercise 5.13

Gage length = ℓ_0 = 50 mm

$E_{mild\ steel}$ = 200 GPa

σ_{yield} = 250 MPa

$\varepsilon_{yield} = \dfrac{250(10^6)}{200(10^9)}$

$= 0.001250 = 1250\ \mu m/m$

Design for ε_{max} = 1500 μm/m

$$\delta = \frac{b}{a} \ell_0 \varepsilon$$

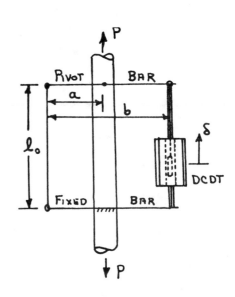

Let b = 2a:

$$\delta_{max} = 2(50)(0.001500) = 0.15\ mm = 0.0059\ in.$$

From Table 5.1, select a DCDT (model 050 DC-D) which has a nominal linear range of ±0.050 in. (1.27 mm) and a scale factor of 200 V/in.

At yield:

$$\delta = 2(50)(0.001250) = 0.125\ mm = 0.00492\ in.$$

$$v_o = 0.00492(200) = 0.984\ V$$

$$S = \frac{v_o}{\varepsilon} = \frac{0.984}{1250} = 0.787(10^{-3})\ V/(\mu in./in.)$$

$$= 0.787\ mV/(\mu in./in.)$$

Limiting the range improves the resolution.

Exercise 5.14

Characteristic	Potentiometer	LVDT
Range	Large	Small
Accuracy	Good	Fair
Resolution	Good	Good
Freq. Response	Poor	Fair
Reliability	Good	Good
Complexity	Simple	Moderate
Cost	Low	Moderate

Exercise 5.15

From Eq. (5.3):
$$\frac{dR}{R} = \frac{d\rho}{\rho} + \frac{dL}{L} - \frac{dA}{A}$$
$$= \frac{d\rho}{\rho} + (1 + 2\nu)\frac{dL}{L}$$

1. Change in specific resistance (ρ).
2. Change in dimensions (A and L)

For constantan: $\quad 1 + 2\nu \approx 1.6$

Therefore: $\quad \frac{1.6}{2.1}(100) = 76.2\%$

Thus, change in dimensions is most important for constantan.

Exercise 5.16

1. Plastic film serves as a carrier for the fragile metal-film grid.
2. Plastic film provides electrical insulation.

Exercise 5.17

From Eq. (5.5): $\dfrac{\Delta R}{R} = S_g \varepsilon$ or $\Delta R = R S_g \varepsilon$

(a)
$$\dfrac{\Delta R}{R} = 2.02(1600)(10^{-6}) = 0.003232$$
$$\Delta R = 120(2.02)(1600)(10^{-6}) = 0.388 \ \Omega$$

(b)
$$\dfrac{\Delta R}{R} = 3.47(650)(10^{-6}) = 0.002256$$
$$\Delta R = 350(3.47)(650)(10^{-6}) = 0.789 \ \Omega$$

(c)
$$\dfrac{\Delta R}{R} = 2.07(650)(10^{-6}) = 0.001346$$
$$\Delta R = 350(2.07)(650)(10^{-6}) = 0.471 \ \Omega$$

(d)
$$\dfrac{\Delta R}{R} = 2.06(200)(10^{-6}) = 0.000412$$
$$\Delta R = 1000(2.06)(200)(10^{-6}) = 0.412 \ \Omega$$

Exercise 5.18

From Eq. (5.7): $v_o = \tfrac{1}{4} v_s S_g \varepsilon$

(a)
$$v_o = \tfrac{1}{4}(2)(2.02)(1600)(10^{-6}) = 0.001616 \ V = 1.62 \ mV$$
$$v_o = \tfrac{1}{4}(4)(2.02)(1600)(10^{-6}) = 0.003232 \ V = 3.23 \ mV$$
$$v_o = \tfrac{1}{4}(7)(2.02)(1600)(10^{-6}) = 0.005656 \ V = 5.66 \ mV$$
$$v_o = \tfrac{1}{4}(10)(2.02)(1600)(10^{-6}) = 0.008080 \ V = 8.08 \ mV$$

(b)
$$v_o = \tfrac{1}{4}(2)(3.47)(650)(10^{-6}) = 0.001128 \ V = 1.13 \ mV$$
$$v_o = \tfrac{1}{4}(4)(3.47)(650)(10^{-6}) = 0.002256 \ V = 2.26 \ mV$$
$$v_o = \tfrac{1}{4}(7)(3.47)(650)(10^{-6}) = 0.003947 \ V = 3.95 \ mV$$
$$v_o = \tfrac{1}{4}(10)(3.47)(650)(10^{-6}) = 0.005639 \ V = 5.64 \ mV$$

Exercise 5.18 (Continued)

(c)
$$v_o = \tfrac{1}{4}(2)(2.07)(650)(10^{-6}) = 0.000673 \text{ V} = 0.67 \text{ mV}$$
$$v_o = \tfrac{1}{4}(4)(2.07)(650)(10^{-6}) = 0.001346 \text{ V} = 1.35 \text{ mV}$$
$$v_o = \tfrac{1}{4}(7)(2.07)(650)(10^{-6}) = 0.002354 \text{ V} = 2.35 \text{ mV}$$
$$v_o = \tfrac{1}{4}(10)(2.07)(650)(10^{-6}) = 0.003364 \text{ V} = 3.36 \text{ mV}$$

(d)
$$v_o = \tfrac{1}{4}(2)(2.06)(200)(10^{-6}) = 0.000206 \text{ V} = 0.21 \text{ mV}$$
$$v_o = \tfrac{1}{4}(4)(2.06)(200)(10^{-6}) = 0.000412 \text{ V} = 0.41 \text{ mV}$$
$$v_o = \tfrac{1}{4}(7)(2.06)(200)(10^{-6}) = 0.000721 \text{ V} = 0.72 \text{ mV}$$
$$v_o = \tfrac{1}{4}(10)(2.06)(200)(10^{-6}) = 0.001030 \text{ V} = 1.03 \text{ mV}$$

Exercise 5.19

If the supply voltage v_s is increased to 50 V the output voltages in Exercise 5.18 become:

(a) $$v_o = \tfrac{1}{4}(50)(2.02)(1600)(10^{-6}) = 0.0404 \text{ V} = 40.4 \text{ mV}$$

(b) $$v_o = \tfrac{1}{4}(50)(3.47)(650)(10^{-6}) = 0.0282 \text{ V} = 28.2 \text{ mV}$$

(c) $$v_o = \tfrac{1}{4}(50)(2.07)(650)(10^{-6}) = 0.0168 \text{ V} = 16.8 \text{ mV}$$

(d) $$v_o = \tfrac{1}{4}(50)(2.06)(200)(10^{-6}) = 0.00515 \text{ V} = 5.15 \text{ mV}$$

The power dissipated by the gages becomes excessive which would cause severe gage heating.

ENGINEERING MEASUREMENTS by J. W. DALLY, W. F. RILEY, AND K. G. McCONNELL

Exercise 5.20

From Eq. (5.7): $v_o = \frac{1}{4} v_s S_g \varepsilon$ or $\varepsilon = \frac{4v_o}{v_s S_g}$

(a) $\varepsilon = \frac{4v_o}{v_s S_g} = \frac{4(1.5)(10^{-3})}{5(2.02)} = 594(10^{-6}) = 594 \; \mu m/m$

(b) $\varepsilon = \frac{4v_o}{v_s S_g} = \frac{4(3.3)(10^{-3})}{5(3.47)} = 761(10^{-6}) = 761 \; \mu m/m$

(c) $\varepsilon = \frac{4v_o}{v_s S_g} = \frac{4(4.8)(10^{-3})}{5(2.07)} = 1855(10^{-6}) = 1855 \; \mu m/m$

(d) $\varepsilon = \frac{4v_o}{v_s S_g} = \frac{4(5.7)(10^{-3})}{5(2.06)} = 2214(10^{-6}) = 2214 \; \mu m/m$

Exercise 5.21

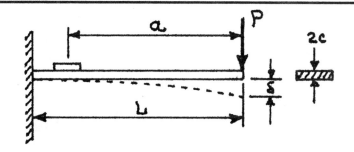

From Eq. (5.7): $v_o = \frac{1}{4} v_s S_g \varepsilon$

Recall that: $\sigma = \frac{Mc}{I}$ and $\delta = \frac{PL^3}{3EI}$

Therefore: $\varepsilon = \frac{\sigma}{E} = \frac{Mc}{EI} = \frac{Pac}{EI} = \frac{3ac\delta}{L^3}$

$$v_o = \frac{1}{4} v_s S_g \frac{3ac\delta}{L^3} = \frac{3ac v_s S_g \delta}{4L^3}$$

$$\delta = \frac{4L^3 v_o}{3ac v_s S_g}$$

Exercise 5.22

From Eq. 2.42:
$$Z_C = -\frac{j}{\omega C}$$

$$Z_C + \Delta Z_C = -\frac{j}{\omega}\left(\frac{1}{C + \Delta C}\right)$$

Dividing by Z_C yields:

$$1 + \frac{\Delta Z_C}{Z_C} = -\frac{j}{\omega}\left(\frac{\omega C}{-j}\right)\left(\frac{1}{C + \Delta C}\right) = \frac{C}{C + \Delta C}$$

$$\frac{\Delta Z_C}{Z_C} = \frac{C}{C + \Delta C} - 1 = -\frac{\Delta C}{C + \Delta C} = -\frac{\Delta C/C}{1 + \Delta C/C}$$

From Eq. 5.9:
$$\frac{\Delta C}{C} = -\frac{\Delta h/h}{1 + \Delta h/h}$$

Thus,
$$\frac{\Delta Z_C}{Z_C} = \frac{\Delta h}{h}$$

From Eqs. 1.6 and 5.8:
$$S = \frac{\Delta Z_C}{\Delta h} = \left|\frac{Z_C}{h}\right| = \left|\frac{1}{\omega C h}\right| = \left|\frac{1}{\omega k K A}\right|$$

Exercise 5.23

From Eq. 5.12:

$S = \dfrac{1}{\omega k K A}$

$K = 1.0$ for air

$k = 0.00885$ for mm dimensions

$A = \dfrac{\pi d^2}{4}$

$1\ pF = 10^{-12}\ F$

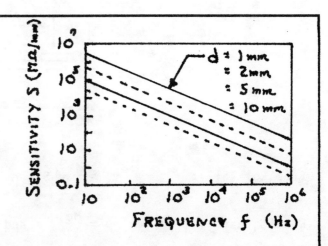

Exercise 5.24

The capacitor sensor has many advantages. It is noncontacting and can be used with a wide variety of materials. The sensor is extremely rugged and can be used in high shock and vibratory environments. Finally, it can be used up to $1600°F$ with excellent stability of its calibration constant. The primary disadvantage is in complexity of the signal conditioning circuits used to measure the changing impedance Z_C from the sensor.

ENGINEERING MEASUREMENTS by J. W. DALLY, W. F. RILEY, AND K. G. McCONNELL

Exercise 5.25

An oscillator-controlled, constant-current ac supply provides the power to the capacitance sensor at a fixed frequency of 15.6 kHz. The voltage across the capacitor changes with the target position and this voltage provides the input to a low-capacitance preamplifier. A second amplifier increases the signal before it is fed into the synchronous detector. The signal is compared (phase) with the digital oscillator, given a sign, and rectified. The rectified signal is filtered to eliminate high frequency noise and to provide an output voltage proportional to Δh. The linearization circuit is used to extend the range of the sensor. Finally, an output amplifier is employed to increase the gain and adjust the sensitivity and zero offset prior to display.

Exercise 5.26

Error will occur.

Output Stainless = 0.82

Output Aluminum = 1.00

$$\%\varepsilon = \frac{1.00 - 0.82}{0.82}(100) = 22.0\%$$

Exercise 5.27

(a) Magnetic materials can be used but at a reduced sensitivity. The sensor must be calibrated for each material.

(b) Polymers are non-conductive and eddy currents cannot be induced in the target. It is necessary to bond a thin aluminum foil to the polymer in the target area.

(c) Non-magnetic metallic foils (like aluminum) are excellent targets. The depth of penetration of the eddy currents is less than 0.001 inch.

Exercise 5.28

From Eq. (5.14):

(a) $q = S_g A p = 2.2(15)(10^{-6})(2)(10^6) = 66$ pC (Quartz)

(b) $q = S_g A p = 130(15)(10^{-6})(2)(10^6) = 3900$ pC (Barium Titanate)

ENGINEERING MEASUREMENTS by J. W. DALLY, W. F. RILEY, AND K. G. McCONNELL

Exercise 5.29

From Eq. (5.16):

(a) $v_o = S_v hp = 0.055(8)(10^{-3})(2)(10^6) = 880$ V (Quartz)

(b) $v_o = S_v hp = 0.011(8)(10^{-3})(2)(10^6) = 176$ V (Barium Titanate)

Exercise 5.30

For the same size and load:

(a) Barium titanate generates more charge (59 to 1). Quartz generates more voltage (5 to 1).

(b) Barium titanate takes less voltage and absorbs more charge; therefore, it is easier to drive in ultrasonic applications.

Exercise 5.31

From Eq. (5.17):

$$\tau_e = \frac{R_p R_A}{R_p + R_A}(C_p + C_L + C_A)$$

For $C_L = 10$ pF:

$$\tau_e = \frac{10(10^{12})(1)(10^9)}{10(10^{12}) + (1)(10^9)}(20 + 10 + 15)(10^{-12})$$

$$= 45.0(10^{-3}) = 45.0 \text{ ms}$$

For $C_L = 1000$ pF:

$$\tau_e = \frac{10(10^{12})(1)(10^9)}{10(10^{12}) + (1)(10^9)}(20 + 1000 + 15)(10^{-12})$$

$$= 1035(10^{-3}) = 1.035 \text{ s}$$

Response to a step input is:

$$e^{-t/\tau_e} = 0.95$$

$$e^{t/\tau_e} = 1/0.95 = 1.0526$$

$$t = \tau_e \ln 1.0526 = 1.035 \ln 1.0526$$

$$= 53.1(10^{-3}) \text{ s} = 53.1 \text{ ms}$$

Exercise 5.32

Piezoelectric: Piezoelectric sensors produce an electric charge when subjected to pressure. Charge decays to zero with time. Piezoelectric sensors have a very high frequency response.

Piezoresistive: Piezoresistive sensors exhibit a resistance change when subjected to pressure. The sensitivity of piezoresistive sensors is very high.

Exercise 5.33

Advantages:

1. High sensitivity (100 versus 2).
2. High resistance (low currents).
3. Transducers with these sensors can be very small.

Disadvantages:

1. Very sensitive to temperature (Thermistors).
2. Can't be applied to curved surfaces; therefore, application as strain gages is limited.

Exercise 5.34

From Fig. 5.18: $i = HA_D$

Therefore, $v_o = iR_L = HA_D R_L$

From Eq. 5.24:

$$R = \frac{v_o}{HA_D} = \frac{HA_D R_L}{HA_D} = R_L$$

For $R_L = 10\ \Omega$: $R = 10$ W

For $R_L = 100\ \Omega$: $R = 100$ W

For $R_L = 1000\ \Omega$: $R = 1000$ W

For $R_L = 10{,}000\ \Omega$: $R = 10{,}000$ W

Exercise 5.35

From Eq. 5.28:
$$S = R\frac{A_D}{R_L}$$

$$R = \frac{SR_L}{A_D} = \frac{50(10^{-3})(100,000)}{8} = 625 \text{ V/W}$$

Exercise 5.36

A photomultiplier tube is a multistage vacuum-tube detector used to measure incident radiation. The tube contains a photocathode which emits electrons in response to incident light (photons). These electrons are collected and focused by circular anodes to produce a beam of accelerated electrons. The beam then proceeds into a series of box and grid dynodes. Interaction of the impinging beam of electrons with the dynodes produces secondary electrons which increase the flow of electrons and the gain in the tube. When the electrons are collected at the anode they produce a current flow on the positive output.

An advantage is the ability to accurately detect small levels of light because of the very high gains (2×10^6).

A disadvantage is the requirement that a constant high voltage must be imposed on the tube in order to maintain stability of the gain. This requirement dictates use of high-quality, well-regulated, high-voltage power supplies which are expensive.

Exercise 5.37

In a photoconducting tube, impinging light drives electrons from a photocathode. As the electrons are collected on an anode they produce a current that is measured.

In a photoconduction cell, the sensitive material is a semiconductor like CdS or CdSe. When a photon interacts with CdS, an electron is driven from the valence band to the conduction band. The result is an increase in charge carriers in the CdS which decreases its resistance. The change in resistance is measured.

Exercise 5.38

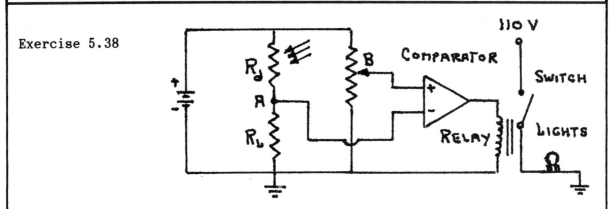

In the sketch above, the voltage at B is adjusted by changing the wiper position on the potentiometer. When the light is intense so that $v_A > v_B$, the comparator output is zero and the relay is open. When the light intensity drops so that $v_B > v_A$, the relay closes and the lights turn on.

Exercise 5.39

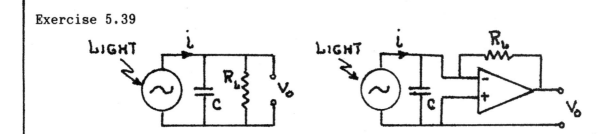

In the simple circuit, the photovoltaic cell acts like a current generator connected to an internal capacitance C and load resistance R_L. If $i = KI$, then the governing differential equation becomes

$$R_L C \dot{v}_o + v_o = R_L KI$$

Thus, the steady state output increases with increasing load resistance R_L; however, the larger $R_L C$ time constant reduces the usable frequency range. The response to a step input light change has the form $(1 - e^{-t R_L C})$. When the cell is connected to an op-amp with an open loop gain G, the governing differential equation becomes

$$\left(\frac{R_L C}{1 + G}\right) \dot{v}_o + V_o = - R_L \left(\frac{G}{1 + G}\right) KI$$

Thus, it is evident that the limiting time constant $R_L C$ is reduced by the open loop gain G of the op-amp so that the photovoltaic cell responds to much high frequencies. The capacitance effect is due to the time it takes the displaced electrons to migrate to the sensing contacts. Essentially, the op-amp shorts the simple circuit. Photovoltaic cells are small, rugged, inexpensive, and have a high sensitivity.

Exercise 5.40

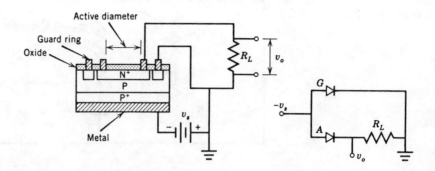

The semiconductor P/N junction shown in the figure above responds to light with a current flow i. The differences in this sensor are the guard ring which adds a second diode to the circuit and the negative bias which is applied to both the active and the guard ring diodes. Current flow through a load resistor produces a voltage which is proportional to the incident radiation. The reverse bias acts to accelerate the electron transition times and this acceleration improves the frequency response of the sensor to about 100 MHz.

Exercise 5.41

For a 100-Ω platinum RTD at 0°C (See Section 5.10):

T (°C)	S (Ω/°C)
0	0.390
100	0.378
200	0.367
300	0.355
400	0.344
500	0.332

A plot of this data shows that it falls on a straight line

From Eq. 1.6:
$$S = \frac{\Delta R}{\Delta T} = R_0[a + b(T - T_0)]$$

or since $T_0 = 0$:
$$S = \frac{\Delta R}{\Delta T} = R_0(a + bT)$$

$$\Delta R = R_0(a + bT)\Delta T$$

$$R = R_0(1 + aT + bT^2)$$

Exercise 5.41 (Continued)

From the above graph and given data :

$i = 10$ mA, $R_0 = 100\ \Omega$, $R_s = 1000\ \Omega$, $a = 3.90(10^{-3})$, and $b = -1.16(10^{-6})$

For Fig. E5.41a: $\quad v_o = iR = iR_0(1 + aT + bT^2)$

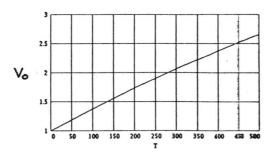

For Fig. E5.41b: $\quad v_o = \dfrac{iR}{1 + R/R_s} = \dfrac{iR_0(1 + aT + bT^2)}{1 + (R_0/R_s)(1 + aT + bT^2)}$

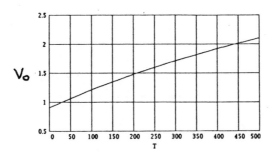

For Fig. E5.41c: $\quad v_o = iR = iR_0(1 + aT + bT^2)$

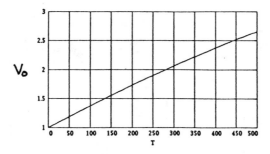

Exercise 5.42

In Figs. 5.41a and c, the output voltage (v = iR) is due to the constant current supply. In Fig. E5.41b the constant current is shared with the shunt resistor. This sharing of current reduces sensitivity and increases non-linear behavior.

Exercise 5.43

Both RTDs and thermistors change resistance with temperature; however, the output ΔR from the thermistor is much larger. Therefore, thermistors should be used to measure small changes in temperature. The output from an RTD is more stable at higher temperatures. Accordingly, RTDs are better suited for high temperature measurements. Most thermistors should not be employed above 500°F.

Exercise 5.44

From Table 5.4:

(a) Chromel-Alumel $S = 25.8 - (-13.6) = 39.4\ \mu V/°C$
(b) Copper-constantan $S = 6.5 - (-35.0) = 41.5\ \mu V/°C$
(c) Iron-constantan $S = 18.5 - (-35.0) = 53.5\ \mu V/°C$
(d) Iron-nickel $S = 18.5 - (-15.0) = 33.5\ \mu V/°C$
(e) Gold-silver $S = 6.5 - (6.5) = 0$

Exercise 5.45

From Eq. 5.33:
$$v_0 = S_{A/B}(T_1 - T_2)$$

Case	Temperature (°C)			Thermocouple combination				
				(a) 39.4 µV	(b) 41.5 µV	(c) 53.5 µV	(d) 33.5 µV	(e) 0 µV
	T_1	T_2	ΔT	Output voltage (mV)				
a	300	0	300	11,820	12,450	16,050	10,050	0
b	200	0	200	7,880	8,300	10,700	6,700	0
c	250	10	240	9,456	9,960	12,840	8,040	0
d	-100	100	-200	-7,880	-8.300	-10,700	-6,700	0

ENGINEERING MEASUREMENTS by J. W. DALLY, W. F. RILEY, AND K. G. McCONNELL

Exercise 5.46

A crystal oscillator utilizes a piezoelectric crystal that is excited at its natural frequency with voltage pulses. Feedback from the crystal maintains the oscillator at its natural frequency for very long periods of time. Short term stability, measured in hours, is about ±1 count in 10^9. Long term stability is ±1 count in 10^5 to 10^8 depending on the quality of the crystal.

Exercise 5.47

1 year = 365 × 24 hr = 8,760 hr = 525,600 min = 31.536×10^6 sec.

Assume aging changes frequency 1 ppm in one year and in addition assume long term stability changes frequency 1 ppm in one year.

Consider a 10 MHz crystal (nominal 10 MHz frequency)

$$f = 10(10^6) \text{ Hz} \quad \text{or} \quad T = 10^{-7} \text{ sec}$$
$$1 \text{ year} = 31.536(10^{13}) \text{ counts}$$
$$\text{error in counts} = 2(10^{-6})(31.536)(10^{13}) = 6.3072(10^8) \text{ counts}$$
$$\text{error per year} = 6.3072(10^8)(10^{-7}) = 63.072 \text{ s}$$

Exercise 5.48

The summary should contain a brief paragraph on each of the following sensors. Each paragraph should indicate the significant advantages or disadvantages based on size, range, accuracy, sensitivity, frequency response, stability, temperature limits, economy, and ease of use.

   (a) Potentiometers
   (b) Differential transformers
   (c) Strain gages (resistive)
   (d) Capacitance sensors
   (e) Eddy-current sensors
   (f) Piezoelectric sensors
   (g) Piezoresistive sensors
   (h) Photoelectric sensors
   (i) Temperature sensors
   (j) Oscillators

ENGINEERING MEASUREMENTS by J. W. DALLY, W. F. RILEY, AND K. G. McCONNELL

Exercise 6.1

When v_i in Fig. 6.1 exceeds v_z in reverse bias, the Zener diode breaks down and the current flow increases rapidly.

From Fig. 6.1b with v_s = 9 V, v_z = 5.2 V, and i = 100 mA:

$$R = \frac{v_s - v_z}{i} = \frac{9 - 5.2}{0.100} = 38 \ \Omega$$

A 1000-Ω load will draw approximately 5 ma, which will cause a slight drop in diode current. This reduction in diode current causes a small drop in diode voltage.

Exercise 6.2

When a small resistor is connected in parallel with the diode, the current draw will lower the diode voltage significantly. This causes the reference voltage to be lost. Thus, Zener diodes can tolerate only high load resistances.

Exercise 6.3

	Battery Type		
Characteristic	LCR	Ni-Cd	Li-I
Power requirements	large	medium	low
recharging	1000 times	500 times	No
range of voltage	4 - 12V	1.2 V/cell	2.75 V/cell
capacity	16 VA·h/lb	16 VA·h/lb	110 VA·h/lb
Discharge rate	20 A/hr	-	-
storage life	> 6 mo	-	long life[*]
Orientation	any	any	any

[*]Lithium batteries exhibit a life of 3 to 9 years with low voltage decay.

Exercise 6.4

Z_o is a complex impedance consisting of a resistance R and an inductance L.

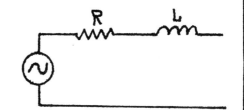

$R = 2 \text{ m}\Omega$
$L = 1 \text{ }\mu\text{H}$ } typical values

Exercise 6.5

From Eq. 6.10:
$$S_{cv} = \frac{r}{1+r}\sqrt{p_T R_T}$$

For $p_T R_T$ = 100, 200, 500, and 1000 W·Ω and $0.1 \le r \le 10$,

	$P_T R_T$ (W·Ω)			
	100	200	500	1000
r	S_{cv}			
0.1	0.91	1.29	2.03	2.87
0.5	3.33	4.71	7.45	10.54
1.0	5.00	7.07	11.18	15.81
2.0	6.67	9.43	14.91	21.08
5.0	8.33	11.79	18.63	26.35
10.0	9.09	12.86	20.33	28.75

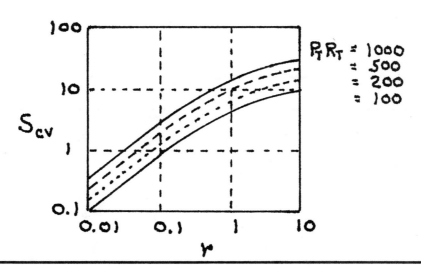

Exercise 6.6

From Eq. (5.5):
$$\frac{\Delta R}{R} = S_g \varepsilon = 2.00(1200)(10^{-6})$$
$$= 2.40(10^{-3}) = 0.0024$$

From Eq. (6.2):
$$\Delta v_o = \frac{[r/(1+r)^2][\Delta R_1/R_1]}{1 + [1/(1+r)][\Delta R_1/R_1]} v_s$$

$$= \frac{[2/(1+2)^2][0.0024]}{1 + [1/(1+2)][0.0024]} (8)$$

$$= 4.26(10^{-3}) \text{ V} = 4.26 \text{ mV}$$

Exercise 6.7

From Eq. (5.5):
$$\frac{\Delta R}{R} = S_g \varepsilon = 2.00(1200)(10^{-6}) = 0.0024$$

From Eq. (6.3):
$$\eta = 1 - \frac{1}{1 + [1/(1+r)][\Delta R_1/R_1]}$$

$$= 1 - \frac{1}{1 + [1/(1+2)][0.0024]}$$

$$= 7.99(10^{-4}) = 0.08\%$$

Exercise 6.8

From Eq. (6.9):
$$v_s = (1 + r)\sqrt{p_g R_g}$$

$$= (1 + 2)\sqrt{(0.25)(350)} = 28.1 \text{ V}$$

Exercise 6.9

From Eq. (6.6):
$$S_{cv} = \frac{r}{(1+r)^2} v_s$$

$$= \frac{2}{(1+2)^2}(8) = 1.778 \text{ V}$$

Exercise 6.10

From Eq. (6.4):

$$\Delta v_o = \frac{r}{(1+r)^2}\left(\frac{\Delta R_1}{R_1} - \frac{\Delta R_2}{R_2}\right)(1-\eta)v_s$$

With no meter, $\Delta R_2 = 0$, and $\eta \cong 0$:

$$\Delta v_o = \frac{r}{(1+r)^2}\left(\Delta R_1/R_1\right)v_s = \frac{R_2 R_g}{(R_g + R_2)^2}\left(\Delta R_g/R_g\right)v_s$$

With a recording instrument in parallel with the gage:

$$R_1 = R_e = \frac{R_g R_m}{R_g + R_m}$$

Therefore:

$$\Delta v_{om} = \frac{R_2 R_e}{(R_e + R_2)^2}\left(\Delta R_e/R_e\right)v_s = \frac{R_2 R_g}{[R_g + R_2(R_g/R_m) + R_2]^2}\left(\Delta R_g/R_g\right)v_s$$

Writing Δv_{om} in terms of Δv_o and a loss factor $\mathcal{L}$ gives

$$\Delta v_{om} = \Delta v_o (1 - \mathcal{L})$$

where

$$\mathcal{L} = \frac{2r(R_g/R_m)\{1 + r[1 + (1/2)(R_g/R_m)]\}}{\{1 + r[1 + (R_g/R_m)]\}^2}$$

With the oscilloscope ($R_m = 10^6 \ \Omega$):

$$R_g/R_m = 350/10^6 = 0.000350$$

$$\mathcal{L} = \frac{2(2)(0.000350)\{1 + 2[1 + 0.5(0.000350)]\}}{[1 + 2(1 + 0.000350)]^2} = 0.000467 = 0.0467\%$$

With the oscillograph ($R_m = 350 \ \Omega$):

$$R_g/R_m = 350/350 = 1.000$$

$$\mathcal{L} = \frac{2(2)(1.000)\{1 + 2[1 + 0.5(1.000)]\}}{[1 + 2(1 + 1.000)]^2} = 0.640 = 64.0\%$$

Exercise 6.11

From Eq. (5.5):
$$\frac{\Delta R}{R} = S_g \varepsilon = 2.00(1200)(10^{-6}) = 0.0024$$

From Eq. (6.12):
$$\Delta v_o = iR\left(\frac{\Delta R}{R}\right) = 0.003(350)(0.0024)$$
$$= 2.52(10^{-3}) \text{ V} = 2.52 \text{ mV}$$

Exercise 6.12

$$\eta = 0$$

Exercise 6.13

$$i = \sqrt{p_g/R_g} = \sqrt{0.25/350} = 26.7(10^{-3}) \text{ A} = 26.7 \text{ mA}$$

Exercise 6.14

From Eq. (6.13):
$$S_{cc} = iR_g = 0.003(350) = 1.05 \text{ V}$$

Exercise 6.15

From Eq. (6.14):
$$S_{cc} = \sqrt{p_g R_g} = \sqrt{(0.25)(350)} = 9.35 \text{ V}$$

Exercise 6.16

From Eq. 6.14: $$S_{cc} = \sqrt{p_T R_T}$$

	Circuit Sensitivity S_{cc}			
R_T	$p_T = 0.1$ W	$p_T = 0.2$ W	$p_T = 0.5$ W	$p_T = 1.0$ W
100	3.16	4.47	7.07	10.00
500	7.07	10.00	15.81	22.36
1000	10.0	14.14	22.36	31.62
5,000	22.36	31.62	50.00	70.71
10,000	31.62	44.72	70.71	100.00

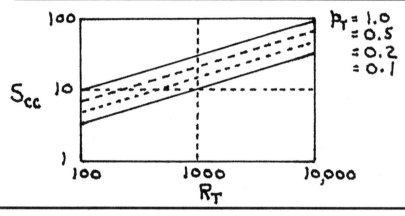

Exercise 6.17

From Eq. (6.21): $$\eta = \frac{\Delta R_p / R_p}{2 + \Delta R_p / R_p} = \frac{\Delta R_p / 1000}{2 + \Delta R_p / 1000} = \frac{\Delta R_p}{2000 + \Delta R_p}$$

ΔR_p	η
-1000	-1.00
-750	-0.60
-500	-0.33
-250	-0.14
0	0
+250	+0.11
+500	+0.20
+750	+0.27
+1000	+0.33

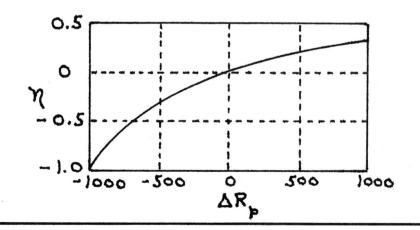

ENGINEERING MEASUREMENTS by J. W. DALLY, W. F. RILEY, AND K. G. McCONNELL

Exercise 6.18

From Eq. (6.19):
$$\Delta v_o = \frac{r}{(1+r)^2}\left(\frac{\Delta R_p}{R_p}\right)(1-\eta)v_s$$

From Exercise 6.17:
$$r = 1, \quad \eta = \frac{\Delta R_p}{2000 + \Delta R_p}$$

$$\Delta v_o = \frac{1}{(1+1)^2}\left(\frac{\Delta R_p}{1000}\right)\left(\frac{2000}{2000 + \Delta R_p}\right)(8) = \frac{4\Delta R_p}{2000 + \Delta R_p}$$

Exercise 6.19

1. The nonlinear effects can be reduced by employing a constant-current source in place of the constant-voltage source. A comparison of Figs. 6.9 and 6.11 shows the improvement.

2. Increasing r also decreases the nonlinear effect as indicated by Eq. (6.31). For example,

    with $\Delta R_p/R_p = 0.5$ and $r = 1$: $\quad \eta = 0.111$

    with $\Delta R_p/R_p = 0.5$ and $r = 5$: $\quad \eta = 0.040$

    with $\Delta R_p/R_p = 0.5$ and $r = 9$: $\quad \eta = 0.024$

3. Finally ΔR_p of the potentiometer of Exercise 6.17 can be effectively reduced by placing a resistor in parallel with potentiometer. This parallel resistor produces $\Delta R_e < \Delta R_p$ which reduces the sensitivity of the circuit.

Exercise 6.20

(a) From Eq. (6.23): $\quad v_s = (1 + r)\sqrt{p_g R_g}$

$$r = \frac{v_s - \sqrt{p_g R_g}}{\sqrt{p_g R_g}} = \frac{28 - \sqrt{0.25(350)}}{\sqrt{0.25(350)}} = 1.993$$

$R_1 = R_g = 350\ \Omega \quad\quad R_2 = rR_g = 1.993(350) = 698\ \Omega$

To maintain the balance condition $R_3 = 698\ \Omega$ and $R_4 = 350\ \Omega$

(b) From Eq. (6.24):
$$S_{cv} = \frac{r}{1+r}\sqrt{p_T R_g} = \frac{1.993}{2.993}\sqrt{0.25(350)} = 6.23\ V$$

Exercise 6.21

From Eq. (5.5): $\quad \dfrac{\Delta R_g}{R_g} = S_g \varepsilon = 2.05(1200)(10^{-6}) = 2460(10^{-6}) = 0.00246$

From Eq. (6.20) with $r = 1.993$:

$$\eta = \dfrac{1}{1 + \dfrac{r+1}{\Delta R_g / R_g}} = \dfrac{1}{1 + \dfrac{2.993}{0.00246}} = 0.00082 \text{ (Negligible)}$$

From Eq. (6.19) with $\eta = 0$:

$$\Delta v_o = \dfrac{r}{(1+r)^2}\left(\dfrac{\Delta R_g}{R_g}\right) v_s = \dfrac{1.993}{(2.993)^2}(0.00246)(28) = 15.32(10^{-3}) \text{ V} = 15.32 \text{ mV}$$

Exercise 6.22

(a) Tension gages in positions 1 & 3
Compression gages in positions 2 & 4

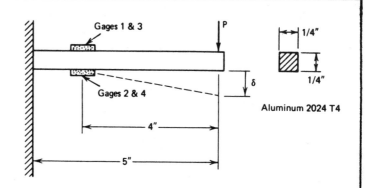

(b) From Eqs. (6.22) and (6.24):

$$S_{cv} = 4\left[\dfrac{r}{1+r}\right]\sqrt{P_T R_g} = 4(1/2)\sqrt{(0.15)(350)} = 14.49 \text{ V}$$

(c) For the cantilever beam: $\quad \sigma = \dfrac{Mc}{I} \quad$ and $\quad \delta = \dfrac{PL^3}{3EI}$

$$\varepsilon = \dfrac{\sigma}{E} = \dfrac{Mc}{EI} = \dfrac{Pac}{EI} = \dfrac{3ac\,\delta}{L^3}$$

$$v_o = S_{cv}(\Delta R_g / R_g) = S_{cv} S_g \varepsilon = S_{cv} S_g \left(\dfrac{3ac\,\delta}{L^3}\right)$$

$$= 14.49(2.00)\left(\dfrac{3(4)(1/8)\delta}{(5)^3}\right) = 0.3550\delta$$

$$C = \dfrac{\delta}{v_o} = \dfrac{\delta}{0.3550\delta} = 2.82 \text{ in./V}$$

Exercise 6.23

From Exercise 6.22:
$$\varepsilon = \frac{\sigma}{E} = \frac{Mc}{EI} = \frac{Pac}{EI}$$

$$v_o = S_{cv}(\Delta R_g/R_g) = S_{cv}S_g\,\varepsilon = S_{cv}S_g\left[\frac{Pac}{EI}\right]$$

With $E = 10\,(10^6)$ psi and $I = \frac{1}{12}\left(\frac{1}{4}\right)^4 = 325.5(10^{-6})$ in.^{4}

$$EI = 10(10^6)(325.5)(10^{-6}) = 3255$$

$$v_o = 14.49(2.00)\left[\frac{P(4)(1/8)}{3255}\right] = 0.004452P$$

$$C = \frac{P}{v_o} = \frac{P}{0.004452P} = 225 \text{ lb/V}$$

Exercise 6.24

From Eq. (6.30):
$$v_o = \frac{i_s R_g r}{2(1+r)}\,\frac{\Delta R_g}{R_g}$$

Eq. (6.30) indicates that v_o increases with r. A practical limit is about 10. Therefore:

$$R_1 = R_4 = 350\ \Omega \qquad R_2 = R_3 = 10(350) = 3500\ \Omega$$

$$i_{max} = \sqrt{P_T/R_g} = \sqrt{\frac{0.10}{120}} = 16.9 \text{ mA}$$

Since 16.9 mA > 10 mA, the power supply limits the current to 10 mA.

From Eq. (6.32):
$$S_{cc} = \frac{i_s R_T r}{2(1+r)} = \frac{10(10^{-3})(350)(10)}{2(1+10)} = 1.591 \text{ V}$$

From Eq. (5.5):
$$\frac{\Delta R_g}{R_g} = S_g\,\varepsilon = 2.00(1500)(10^{-6})$$
$$= 3000(10^{-6}) = 0.0030$$

$$v_o = \frac{i_s R_g r}{2(1+r)}\,\frac{\Delta R_g}{R_g} = \frac{10(10^{-3})(350)(10)}{2(1+10)}(0.0030)$$
$$= 4.77(10^{-3}) \text{ V} = 4.77 \text{ mV}$$

Exercise 6.25

From Eq. (6.31):

$$\eta = \frac{\Delta R_p/R_p}{2(1+r) + \Delta R_p/R_p}$$

$$= \frac{\Delta R_p/1000}{4 + \Delta R_p/1000} = \frac{\Delta R_p}{4000 + \Delta R_p}$$

ΔR_p	η
-1000	-0.333
-750	-0.231
-500	-0.143
-250	-0.067
0	0
+250	+0.059
+500	+0.111
+750	+0.158
+1000	+0.200

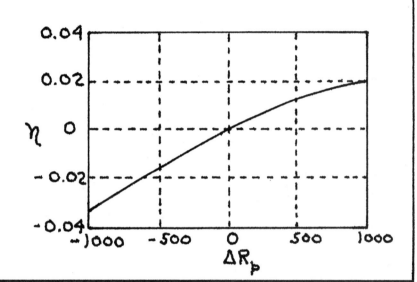

Exercise 6.26

From Eq. (6.30):

$$v_o = \frac{i_s R_p r}{2(1+r)} (1 - \eta) \frac{\Delta R_p}{R_p}$$

From Exercise 6.25:

$$1 - \eta = 1 - \frac{\Delta R_p}{4000 + \Delta R_p} = \frac{4000}{4000 + \Delta R_p}$$

$$v_o = \frac{20(10^{-3})(1000)(1)}{2(1+1)} \left(\frac{4000}{4000 + \Delta R_p}\right) \left(\frac{\Delta R_p}{1000}\right)$$

$$= \frac{20 \, \Delta R_p}{4000 + \Delta R_p}$$

Exercise 6.27

From Eq. 6.36:
$$\frac{v_o}{v_i} = G(1 - e^{-t/\tau})$$

Voltage ratio v_o/v_i

t (µs)	Gain		
	10	100	1000
0	0	0	0
10	6.32	63.2	632
20	8.65	86.5	865
50	9.93	99.3	993
100	9.99	99.9	999

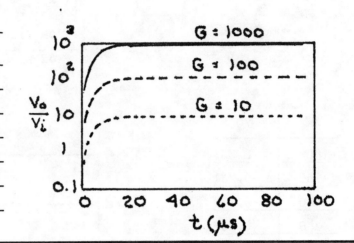

Exercise 6.28

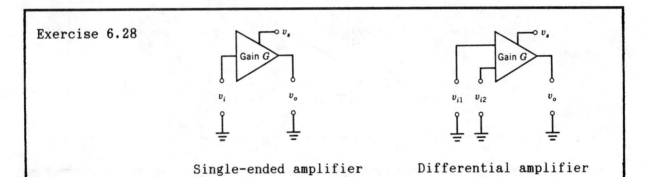

Single-ended amplifier   Differential amplifier

Exercise 6.29

From Eq. 6.38:
$$v_o = G_d \Delta v + G_c v$$

From Eq. 6.39:
$$G_c = \frac{G_d}{\text{CMRR}} = \frac{500}{\text{CMRR}}$$

With $v = 0.10$ V and $\Delta v = 10$ mV:

CMRR	$G_d \Delta v$	G_c	$G_c v$	v_o
1000	5.000	0.500	0.0500	5.0500
5000	5.000	0.100	0.0100	5.0100
10,000	5.000	0.050	0.0050	5.0050
20,000	5.000	0.025	0.0025	5.0025

Exercise 6.30

From Eq. 6.44: $\quad R_d = 20 \log_{10}(v_m/v_n) \qquad v_n = v_{nA}/G$

Therefore: $\quad R_d = 20 \log\left(\dfrac{0.500}{5(10^{-6})/G}\right)$

$\qquad\qquad\qquad = 20 \log(100{,}000\,G) = 100 + 20 \log G$

G	R_d
1	100
10	120
50	134
100	140
500	154
1000	160

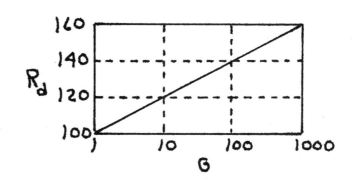

Exercise 6.31

From Eq. (3.10): $\quad G = 100\text{ dB} = 10^5$

From Eq. (6.45): $\quad G_c = \dfrac{v_o}{v_i} = -\dfrac{R_f}{R_1}\left\{\dfrac{1}{1 + (1/G)[1 + (R_f/R_1) + (R_f/R_a)]}\right\}$

(a) For $G = 10$ use $R_f/R_1 = 10$: $\qquad R_f = 1\text{ M}\Omega$ and $R_1 = 100\text{ k}\Omega$

$$G_c = -10\left\{\dfrac{1}{1 + (1/10^5)[1 + 10 + (1/7)]}\right\} = -9.999 \cong -10$$

(b) For $G = 20$ use $R_f/R_1 = 20$: $\qquad R_f = 1\text{ M}\Omega$ and $R_1 = 50\text{ k}\Omega$

$$G_c = -20\left\{\dfrac{1}{1 + (1/10^5)[1 + 20 + (1/7)]}\right\} = -19.9958 \cong -20$$

(c) For $G = 50$ use $R_f/R_1 = 50$: $\qquad R_f = 1\text{ M}\Omega$ and $R_1 = 20\text{ k}\Omega$

$$G_c = -50\left\{\dfrac{1}{1 + (1/10^5)[1 + 50 + (1/7)]}\right\} = -49.974 \cong -50$$

(d) For $G = 100$ use $R_f/R_1 = 100$: $\qquad R_f = 1\text{ M}\Omega$ and $R_1 = 10\text{ k}\Omega$

$$G_c = -100\left\{\dfrac{1}{1 + (1/10^5)[1 + 100 + (1/7)]}\right\} = -99.899 \cong -100$$

Exercise 6.32

From Eq. (3.10): $\quad G = 100 \text{ dB} = 10^5$

From Eq. (6.47):

$$G_c = \frac{v_o}{v_i} = \frac{G}{1 + GR_1/(R_1 + R_f)} = \frac{G(R_1 + R_f)}{(R_1 + R_f) + GR_1} = \frac{1 + R_f/R_1}{(1 + R_f/R_1)/G + 1}$$

(a) For $G = 10$ use $R_f/R_1 = 9$: $\quad R_f = 900 \text{ k}\Omega$ and $R_1 = 100 \text{ k}\Omega$

$$G_c = \frac{1 + R_f/R_1}{(1 + R_f/R_1)/G + 1} = \frac{1 + 9}{(1 + 9)/10^5 + 1} = 9.9990 \cong 10$$

(b) For $G = 20$ use $R_f/R_1 = 19$: $\quad R_f = 1.9 \text{ M}\Omega$ and $R_1 = 100 \text{ k}\Omega$

$$G_c = \frac{1 + R_f/R_1}{(1 + R_f/R_1)/G + 1} = \frac{1 + 19}{(1 + 19)/10^5 + 1} = 19.9960 \cong 20$$

(c) For $G = 50$ use $R_f/R_1 = 49$: $\quad R_f = 4.9 \text{ M}\Omega$ and $R_1 = 100 \text{ k}\Omega$

$$G_c = \frac{1 + R_f/R_1}{(1 + R_f/R_1)/G + 1} = \frac{1 + 49}{(1 + 49)/10^5 + 1} = 49.9750 \cong 50$$

(d) For $G = 100$ use $R_f/R_1 = 99$: $\quad R_f = 9.9 \text{ M}\Omega$ and $R_1 = 100 \text{ k}\Omega$

$$G_c = \frac{1 + R_f/R_1}{(1 + R_f/R_1)/G + 1} = \frac{1 + 99}{(1 + 99)/10^5 + 1} = 99.9000 \cong 100$$

Exercise 6.33

From Eq. (3.10): $\quad G = 100 \text{ dB} = 10^5$

From Eq. (6.50): $\quad G_c \cong \dfrac{R_f}{R_1}$

(a) For $G = 10$ use $R_f/R_1 = 10$: $\quad R_f = 1 \text{ M}\Omega$ and $R_1 = 100 \text{ k}\Omega$

(b) For $G = 20$ use $R_f/R_1 = 20$: $\quad R_f = 1 \text{ M}\Omega$ and $R_1 = 50 \text{ k}\Omega$

(c) For $G = 50$ use $R_f/R_1 = 50$: $\quad R_f = 1 \text{ M}\Omega$ and $R_1 = 20 \text{ k}\Omega$

(d) For $G = 100$ use $R_f/R_1 = 100$: $\quad R_f = 1 \text{ M}\Omega$ and $R_1 = 10 \text{ k}\Omega$

Exercise 6.34

From Eq. (3.10): $\quad G = 120 \text{ dB} = 10^6$

From Eq. (6.52): $\quad R_{ci} = (1 + G) R_a = (1 + 10^6)(10)(10^6) = 10(10^{12}) \; \Omega$

R_0 is typically 100 Ω for an op-amp.

From Eq. (6.53): $\quad R_{co} = \dfrac{R_0}{1 + G} = \dfrac{100}{1 + 10^6} = 100(10^{-6}) \; \Omega = 100 \; \mu\Omega$

Exercise 6.35

From Eq. 6.37:
$$v_o = G(v_{i1} - v_{i2}) = G(v_i - v_o)$$

$$v_o = \frac{G}{1 + G} v_i \qquad i_a = \frac{v_i - v_o}{R_a}$$

$$R_{ci} = \frac{v_i R_a}{v_i - v_o} = \frac{v_i R_a}{v_i - \frac{G}{1 + G} v_i} = (1 + G) R_a$$

Exercise 6.36

At Point A.

$i_1 + i_2 + i_3 + i_f = i_a$

$i_1 = \dfrac{v_{i1} - v_A}{R_1}$

$i_2 = \dfrac{v_{i2} - v_A}{R_2}$

$i_3 = \dfrac{v_{i3} - v_A}{R_3}$

$i_f = \dfrac{v_o - v_A}{R_f}$

$v_o = -G v_A \qquad i_a = \dfrac{v_A}{R_a}$

Exercise 6.36 (Continued)

$$\frac{v_{i1}}{R_1} + \frac{v_{i2}}{R_2} + \frac{v_{i3}}{R_3} + \frac{v_o}{R_f} = \frac{v_A}{R_a} + \frac{v_A}{R_1} + \frac{v_A}{R_2} + \frac{v_A}{R_3} + \frac{v_A}{R_f}$$

$$v_o \left[\frac{1}{R_f} + \frac{1}{G}\left(\frac{1}{R_a} + \frac{1}{R_1} + \frac{1}{R_2} + \frac{1}{R_3} + \frac{1}{R_f}\right)\right] = -\left(\frac{v_{i1}}{R_1} + \frac{v_{i2}}{R_2} + \frac{v_{i3}}{R_3}\right)$$

$$v_o = -\frac{\dfrac{v_{i1}}{R_1} + \dfrac{v_{i2}}{R_2} + \dfrac{v_{i3}}{R_3}}{\dfrac{1}{R_f} + \dfrac{1}{G}\left(\dfrac{1}{R_1} + \dfrac{1}{R_2} + \dfrac{1}{R_3} + \dfrac{1}{R_f} + \dfrac{1}{R_a}\right)}$$

Exercise 6.37

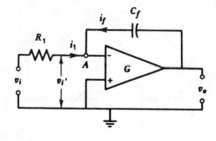

$$i_1 = \frac{v_1 - v_A}{R_1}$$

$$q_f = C(v_o - v_A)$$

$$i_f = \dot{q}_f = C(\dot{v}_o - \dot{v}_A)$$

At Point A:

$$i_i + i_f = \frac{v_i - v_A}{R_1} + C(\dot{v}_o - \dot{v}_A) = 0$$

$$v_o = -G v_A$$

$$\frac{v_i}{R_1} + \frac{v_o}{GR_1} + C\left(\dot{v}_o + \frac{\dot{v}_o}{G}\right) = \frac{v_i}{R_1} + C\dot{v}_o\left(1 + \frac{1}{G}\right) = 0$$

Which yields

$$\dot{v}_o = -\frac{1}{R_1 C} v_i$$

$$v_o = -\frac{1}{R_1 C} \int_0^t v_i \, dt$$

Exercise 6.38

Assume error term in Eq. (6.56) is small.

Then:
$$v_o = -R_f \left[\frac{v_{i1}}{R_1} + \frac{v_{i2}}{R_2} + \frac{v_{i3}}{R_3} \right]$$

With $R_1 = R_f$, $R_2 = \frac{1}{3} R_f$, and $R_3 = 3 R_f$:

$$v_o = -\left(v_{i1} + 3v_{i2} + \frac{1}{3} v_{i3} \right)$$

If input circuit loading is not a problem, use:

$R_f = 30$ kΩ $\qquad R_1 = 30$ kΩ $\qquad R_2 = 10$ kΩ $\qquad R_3 = 90$ kΩ

Exercise 6.39

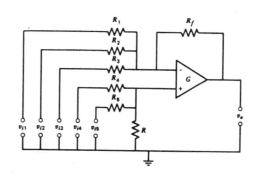

For inputs 1, 2, and 3:

$$i_1 + i_2 + i_3 = i_f$$

$$\frac{v_{i1} - v_A}{R_1} + \frac{v_{i2} - v_A}{R_2} + \frac{v_{i3} - v_A}{R_3} = \frac{v_A - v_o}{R_f}$$

$$\frac{v_{i1}}{R_1} + \frac{v_{i2}}{R_2} + \frac{v_{i3}}{R_3} + \frac{v_o}{R_f} = \frac{v_A}{R_1} + \frac{v_A}{R_2} + \frac{v_A}{R_3} + \frac{v_A}{R_f} = \frac{v_A}{R_A}$$

Where:
$$\frac{1}{R_A} = \frac{1}{R_1} + \frac{1}{R_2} + \frac{1}{R_3} + \frac{1}{R_f}$$

Therefore:
$$v_A = R_A \left(\frac{v_{i1}}{R_1} + \frac{v_{i2}}{R_2} + \frac{v_{i3}}{R_3} + \frac{v_o}{R_f} \right)$$

Exercise 6.39 (Continued)

For inputs 4 and 5:
$$i_4 + i_5 = i$$

$$\frac{v_{i4} - v_B}{R_4} + \frac{v_{i5} - v_B}{R_5} = \frac{v_B}{R}$$

$$\frac{v_{i4}}{R_4} + \frac{v_{i5}}{R_5} = \frac{v_B}{R} + \frac{v_B}{R_4} + \frac{v_B}{R_5} = \frac{v_B}{R_B}$$

Where:
$$\frac{1}{R_B} = \frac{1}{R} + \frac{1}{R_4} + \frac{1}{R_5}$$

Therefore:
$$v_B = R_B\left(\frac{v_{i4}}{R_4} + \frac{v_{i5}}{R_5}\right)$$

$$v_o = G(v_B - v_A)$$

$$= G\left[R_B\left(\frac{v_{i4}}{R_4} + \frac{v_{i5}}{R_5}\right) - R_A\left(\frac{v_{i1}}{R_1} + \frac{v_{i2}}{R_2} + \frac{v_{i3}}{R_3} + \frac{v_o}{R_f}\right)\right]$$

$$= GR_B\left(\frac{v_{i4}}{R_4} + \frac{v_{i5}}{R_5}\right) - GR_A\left(\frac{v_{i1}}{R_1} + \frac{v_{i2}}{R_2} + \frac{v_{i3}}{R_3}\right) - GR_A\frac{v_o}{R_f}$$

$$v_o\left(1 + G\frac{R_A}{R_f}\right) = G\left[R_B\left(\frac{v_{i4}}{R_4} + \frac{v_{i5}}{R_5}\right) - R_A\left(\frac{v_{i1}}{R_1} + \frac{v_{i2}}{R_2} + \frac{v_{i3}}{R_3}\right)\right]$$

$$v_o = \frac{R_f}{R_A}\left[R_B\left(\frac{v_{i4}}{R_4} + \frac{v_{i5}}{R_5}\right) - R_A\left(\frac{v_{i1}}{R_1} + \frac{v_{i2}}{R_2} + \frac{v_{i3}}{R_3}\right)\right]$$

$$v_o \cong \frac{R_f^*}{R_4}v_{i4} + \frac{R_f^*}{R_5}v_{i5} - \frac{R_f}{R_1}v_{i1} - \frac{R_f}{R_2}v_{i2} + \frac{R_f}{R_3}v_{i3}$$

Where:
$$R_f^* = R_f\frac{R_B}{R_A}$$

Exercise 6.40

From Eq. (6.57):

$$v_o = -\frac{1}{R_1 C_f} \int_0^t v_i \, dt$$

$$= -\frac{1}{(10^6)(0.5)(10^{-6})} \int_0^t v_i \, dt = -2 \int_0^t v_i \, dt$$

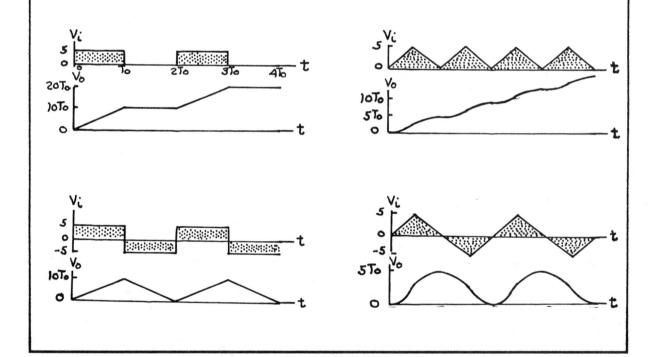

Exercise 6.41

Signals (a) and (c) will increase without bound; and hence, will eventually saturate the amplifier.

Exercise 6.42

From Eq. (6.58):
$$v_o = -R_f C_1 \frac{dv_i}{dt}$$
$$= (10^6)(0.5)(10^{-6}) \frac{dv_i}{dt} = 0.5 \frac{dv_i}{dt}$$

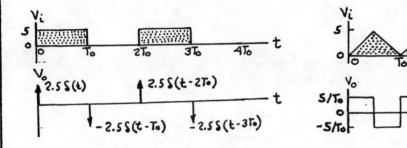
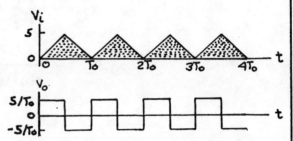
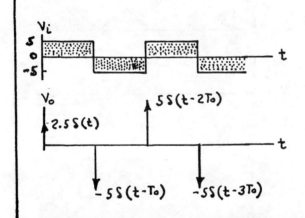
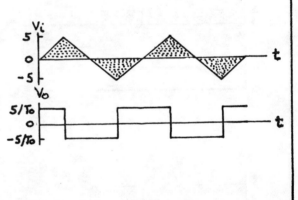

Exercise 6.43

High-pass RC filter:

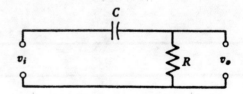

Low-pass RC filter:

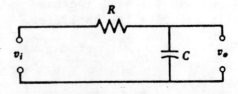

Exercise 6.44

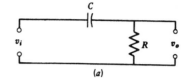

$q = C(v_i - v_o)$

$i = C(\dot{v}_i - \dot{v}_o)$

$v_o = iR$

Therefore:
$$\frac{v_o}{R} + \dot{v}_o = C\dot{v}_i$$

or
$$RC\dot{v}_o + v_o = RC\dot{v}_i \qquad (a)$$

For an input: $\quad v_i = V_i e^{j\omega t}$

For an output: $\quad v_o = V_o e^{j\omega t}$

Substituting in Eq. (a): $\quad (jRC\omega + 1) V_o e^{j\omega t} = jRC\omega V_i e^{j\omega t}$

$$\frac{V_o}{V_i} = \frac{jRC\omega}{1 + jRC\omega} = \frac{RC\omega}{\sqrt{1 + (RC\omega)^2}} e^{j\phi}$$

$$|H(\omega)| = \frac{RC\omega}{\sqrt{1 + (RC\omega)^2}} \qquad \phi = \tan^{-1} RC\omega$$

Exercise 6.45

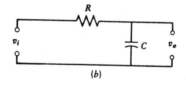

$i = \dfrac{v_i - v_o}{R} = C\dot{v}_o$

$RC\dot{v}_o + v_o = v_i \qquad (a)$

For an input: $\quad v_i = V_i e^{j\omega t}$

For an output: $\quad v_o = V_o e^{j\omega t}$

Substituting in Eq. (a): $\quad (jRC\omega + 1) V_o e^{j\omega t} = V_i e^{j\omega t}$

$$\frac{V_o}{V_i} = \frac{1}{1 + jRC\omega} = \frac{1}{\sqrt{1 + (RC\omega)^2}} e^{-j\phi}$$

$$|H(\omega)| = \frac{1}{\sqrt{1 + (RC\omega)^2}} \qquad \phi = \tan^{-1} RC\omega$$

Exercise 6.46

From Eq. (6.59): $\quad \dfrac{v_o}{v_i} = \dfrac{u}{\sqrt{1 + u^2}} \quad$ where $u = \omega RC$

For 2% Attenuation: $\quad \dfrac{v_o}{v_i} = 0.98 = \dfrac{u}{\sqrt{1 + u^2}} \quad u = 4.925$

$$u = \omega RC = 2\pi f RC = 4.925 \qquad RC = \dfrac{4.925}{2\pi f}$$

(a) For $f = 10$ Hz: $\quad RC = 0.0784 \quad$ If $C = 0.1\ \mu F \quad R = 784\ k\Omega$

(b) For $f = 20$ Hz: $\quad RC = 0.0392 \quad$ If $C = 0.1\ \mu F \quad R = 392\ k\Omega$

(c) For $f = 30$ Hz: $\quad RC = 0.0261 \quad$ If $C = 0.1\ \mu F \quad R = 261\ k\Omega$

Exercise 6.47

From Eq. (6.60): $\quad \dfrac{v_o}{v_i} = \dfrac{1}{\sqrt{1 + u^2}} \quad$ where $u = \omega RC$

For 1% Attenuation: $\quad \dfrac{v_o}{v_i} = 0.99 = \dfrac{1}{\sqrt{1 + u^2}} \quad u = 0.1425$

$$u = \omega RC = 2\pi f RC = 0.1425 \qquad RC = \dfrac{0.1425}{2\pi f}$$

(a) For $f = 5$ Hz: $\quad RC = 0.004536 \quad$ If $C = 1.0\ \mu F \quad R = 4.536\ k\Omega$

$\quad f = 60$ Hz: $\quad RT = \dfrac{1}{\sqrt{1 + u^2}} = \dfrac{1}{\sqrt{1 + (1.71)^2}} = 0.505$ (49% attenuation)

(b) For $f = 10$ Hz: $\quad RC = 0.002268 \quad$ If $C = 1.0\ \mu F \quad R = 2.268\ k\Omega$

$\quad f = 60$ Hz: $\quad RT = \dfrac{1}{\sqrt{1 + u^2}} = \dfrac{1}{\sqrt{1 + (0.855)^2}} = 0.760$ (24% attenuation)

(c) For $f = 20$ Hz: $\quad RC = 0.001134 \quad$ If $C = 1.0\ \mu F \quad R = 1.134\ k\Omega$

$\quad f = 60$ Hz: $\quad RT = \dfrac{1}{\sqrt{1 + u^2}} = \dfrac{1}{\sqrt{1 + (0.428)^2}} = 0.919$ (8% Attenuation

ENGINEERING MEASUREMENTS by J. W. DALLY, W. F. RILEY, AND K. G. McCONNELL

Exercise 6.48

For the notch filter: $\quad f = \dfrac{1}{2\pi RC} \quad$ or $\quad RC = \dfrac{1}{2\pi f}$

(a) For f = 60 Hz: $\quad$ RC = 0.00265 $\quad$ If C = 0.1 μF $\quad$ R = 26.5 kΩ

(b) For f = 1200 Hz: $\quad$ RC = 0.000133 $\quad$ If C = 0.01 μF $\quad$ R = 13.3 kΩ

(c) For f = 10,000 Hz: $\quad$ RC = 0.0000159 $\quad$ If C = 0.001 μF $\quad$ R = 15.9 kΩ

Exercise 6.49

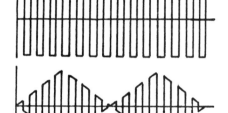

Signal:

Carrier:

Amplitude Modulated Signal:

Exercise 6.50

In many cases it is advantageous to mix the signal from a sensor with that of a carrier to better transmit data. An example of this is in space where sensor signals are transmitted to the earth by telemetry.

One form of signal mixing involves the product of the two signals. This multiplication of the signals produces a new signal that is referred to as being amplitude modulated, because the amplitude of the carrier signal is modulated by the amplitude of the sensor signal as shown in Fig. 6.30.

When the amplitude modulated signal is received, the two components must be separated. This separation process is called demodulation. Demodulation is accomplished by rectifying and filtering the mixed signal. The filter eliminates the high frequency carrier signal while maintaining the sensor signals without attenuation.

Exercise 6.51

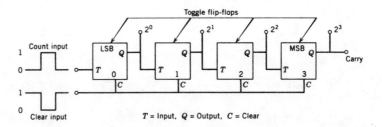

A binary counting unit contains a series of interconnected flip flops. A flip flop stores a bit either high 1 or low 0. The flip flop can be toggled from 0 to 1 to 0, etc. due to a signal on the input (T). Counting begins by clearing all of the flip flops so the output Q for the four-bit counter in Fig. 6.32 is 0000. The count input is a series of pulses that is fed into the LSB flip flop. On the first pulse, the first (LSB) flip flop goes high (from 0 to 1) giving the count 0001. On the second pulse, the LSB flip flop goes low (from 1 to 0), the second flip flop goes high (from 0 to 1), and the register reads 0010 or 2. The process continues until the register is full at a reading of 1111 which corresponds to a count of 15 pulses.

Exercise 6.52

A Schmitt trigger is an electronic circuit which provides either a high or low output depending upon the input signal. When the input signal is below the lower threshold limit, the output is low: conversely, when the input is above an upper threshold limit, the output is high.

The Schmitt trigger is important in counting circuits because it can be used to convert an analog waveform into a train of square-wave pulses that can be counted with a binary counting unit.

Exercise 6.53

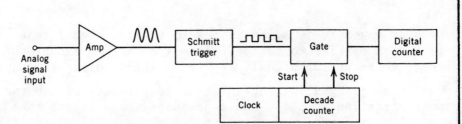

The input signal to the counter (an analog wave) is first amplified and then converted to a train of pulses by a Schmitt trigger. These pulses are transmitted through a gate to a binary counter. A decade counter (driven by a clock) controls the gate and determines the interval of time over which the pulses are counted.

Exercise 6.54

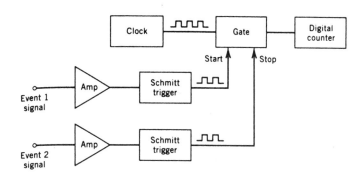

The TIM arrangement is used to measure the time between two events. This measurement is accomplished with a clock, gate, and counter as shown in the sketch. The signals to the gate, which start and stop the count, are due to the events 1 and 2. Amplifiers and Schmitt triggers are used to generate a square wave to switch the gate.

Exercise 6.55

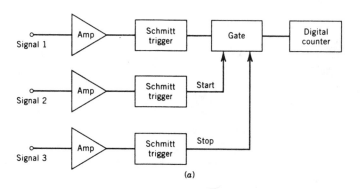

(a)

The GATE function involves counting pulses associated with event 1 during a time interval started by event 2 and terminated by event 3. The analog signal from event 1 is amplified and converted to a train of pulses by a Schmitt trigger. The transmission of this train of pulses to a binary counter is controlled by a gate. The analog signal from event 2, after amplification and squaring, turns the gate on and starts transmission of the train of pulses due to signal 1. The analog signal from event 3, after conditioning like signal 2, is used to turn the gate off.

Exercise 6.56

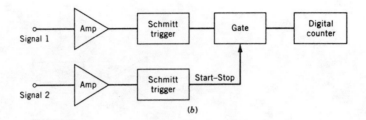
(b)

RATIO is the measurement of the ratio of two frequencies. Signal 1, through the first amplifier/trigger, provides the pulses that are counted. Signal 2 through the second amplifier/trigger, provides the start and the stop pulses for the gate. The binary counter counts the pulses in signal 1 during a single cycle of signal 2 and gives the ratio of A/B where A is the frequency of signal 1 and B is the frequency of signal 2.

Exercise 6.57

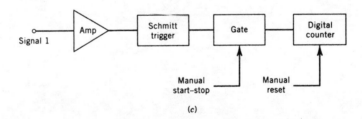
(c)

The count function is completely manual like a digital stop watch. The analog signal is conditioned to provide a train of square pulses by an amplifier and a Schmitt trigger. The gate controls the transmission of the pulses to the binary counter. The gate is started and stopped by manual action.

ENGINEERING MEASUREMENTS by J. W. DALLY, W. F. RILEY, AND K. G. McCONNELL

Exercise 7.1

Given: $d = 0.001$ in., $L_g = 2.00$ in., $R_g = 500\ \Omega$, and $R_w = 25\ \Omega/\text{in.}$

$$L_w = \frac{R_g}{R_w} = \frac{500}{25} = 20.0 \text{ in.}$$

Number of segments $= \dfrac{L_w}{L_g} = \dfrac{20.0}{2.00} = 10$

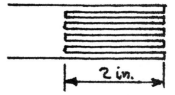

Exercise 7.2

Strain is a point quantity and this fact requires that the gage length and width be small so that error due to nonlinear strain gradients is minimized. The calibration constant that relates output to strain and the output itself should be constant with time, temperature or other environmental factors that may change during the period of measurement. The gage should be sufficiently sensitive and accurate to measure strains to ± 1 μin./in. and have sufficient range to permit measurements of strains as large as ± 10 percent. The output should also be linear over this range. The gage should be sufficiently responsive to permit recording of dynamic strains with frequencies as high as 100 kHz. The gage and related instrumentation should be easy to install and operate, and should permit on-location and remote readout. The costs of the gage, installation, and recording instruments should be low to permit wide spread usage. Finally, the gage should be suitable for use as a sensing element in other transducer systems.

Exercise 7.3

A foil-type electrical resistance strain gage consists of a thin (~100 μin.) metal foil that is supported on a thin (~0.001 in.) plastic carrier. The gage element is produced by photolithography and etching the metal foil. The gage factor and other gage characteristics are determined by the alloys used in the foil elements, which include:

1. Advance for general purpose gages.

2. Karma for fatigue applications and large temperature variations.

3. Isoelastic for high output under dynamic conditions.

The plastic films serve to carry the thin, fragile, metal grids during installation and to electrically insulate the foils from the specimen.

Exercise 7.4

The specification should include the following paragraphs:
1. Surface preparation including cleaning, sanding, and degreasing.
2. Marking the gage location and orientation.
3. Selection of the gage.
4. Positioning the gage -- tape method.
5. Selection of the adhesive.
6. Chemical preparation of the surface for the adhesive.
7. Gage bonding and curing procedure.
8. Gage testing prior to wiring.
9. Terminal tab selection and bonding.
10. Gage wiring.
11. Log record of gage position and characteristics R_g, S_g, and K_t.

Exercise 7.5

The specification should include the following paragraphs:
1. Surface preparation including cleaning, sanding, and degreasing.
2. Marking the gage location and orientation.
3. Selection of the gage. Gage size is very important; both the gage length ℓ and the gage width w should be as large as possible.
4. Positioning the gage -- tape method.
5. Selection of the adhesive.
6. Chemical preparation of the surface for the adhesive.
7. Gage bonding and curing procedure.
8. Gage testing prior to wiring.
9. Terminal tab selection and bonding.
10. Gage wiring.
11. Log record of gage position and characteristics R_g, S_g, and K_t.
12. Warning on maximum bridge voltage that can be used in testing.

Exercise 7.6

From Eq. (7.2):
$$S_s = \frac{r}{1 + r} S_g \sqrt{p_g R_g}$$

With $S_g = 2$, and $r = 3$:

$$S_s = \frac{3}{1 + 3}(2)\sqrt{p_g R_g} = 1.50\sqrt{p_g R_g}$$

	R_g (Ω)			
p_g (W/in.^{2})	120	350	500	1000
0.01	1.64	2.81	3.35	4.74
0.10	5.20	8.87	10.61	15.00
1.00	16.43	28.06	33.54	47.43
2.00	23.24	39.69	47.43	67.08
5.00	36.74	62.75	75.00	106.07
10.00	51.96	88.74	106.07	150.00

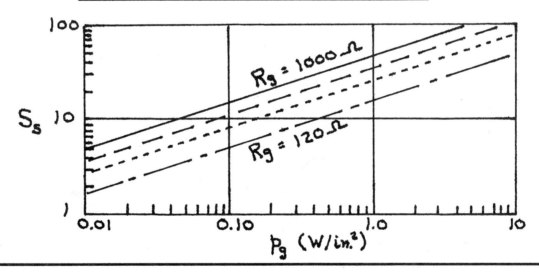

Exercise 7.7

From Eq. (7.1):
$$S_s = S_g S_c = \frac{\Delta v_o}{\varepsilon}$$

From Eq. (6.22):
$$S_c = \frac{r}{(1 + r)^2} v_s$$

$$S_s = S_g \frac{r}{(1 + r)^2} v_s = 2.05\left(\frac{3}{(1 + 3)^2}\right)(5) = 1.922 \; \mu V/(\mu in./in.)$$

Exercise 7.8

From Eq. (7.3):
$$p_g = A\, p_D$$

From Eq. (7.5):
$$v_s = (1 + r)\sqrt{A p_D R_g}$$
$$= (1 + r)\sqrt{p_g R_g}$$
$$= (1 + 3)\sqrt{0.01(350)} = 7.48 \text{ V}$$

The bridge voltage ($v_s = 5$ V) is satisfactory; could use as much as $v_s = 7.48$ V.

Exercise 7.9

From Eq. (6.18):
$$v_o = \frac{r}{(1 + r)^2}\left(\frac{\Delta R_g}{R_g}\right) v_s$$

From Eq. (5.5):
$$\frac{\Delta R_g}{R_g} = S_g \varepsilon$$

Therefore:
$$v_o = \frac{r}{(1 + r)^2} S_g v_s \varepsilon$$
$$= \frac{1}{(1 + 1)^2} (2.06)(6)(1200)(10^{-6})$$
$$= 3708(10^{-6}) \text{ V} = 3.71 \text{ mV}$$

Exercise 7.10

From Eq. (7.5): $v_s = (1 + r)\sqrt{Ap_D R_g}$

With $r_g = 1$ and $R_g = 350\ \Omega$:

$$v_s = (1 + 1)\sqrt{350\ Ap_D} = 37.42\sqrt{Ap_D}$$

	p_D (W/in.^{2})						
A (in.^{2})	0.1	0.2	0.5	1.0	2.0	5.0	10.0
0.001	0.37	0.53	0.84	1.18	1.67	2.65	3.74
0.005	0.84	1.18	1.87	2.65	3.74	5.92	8.37
0.010	1.18	1.67	2.65	3.74	5.29	8.37	11.83
0.050	2.65	3.74	5.92	8.37	11.83	18.71	26.46
0.500	8.37	11.83	18.71	26.46	37.42	59.17	83.67

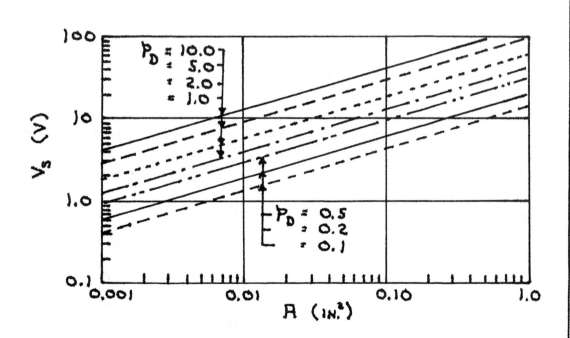

Exercise 7.11

From Eqs. (5.5) and (6.18):

$$v_o = \frac{r}{(1+r)^2} v_s S_g G \varepsilon$$

With $r = 1$, $S_g = 2.07$, and $\varepsilon = 1600$ μin./in.:

$$v_o = \frac{1}{(1+1)^2}(2.07)(1600)(10^{-6}) v_s G = 828(10^{-6}) v_s G$$

Case	v_s (V)	G	v_o (mV)	reading
(a)	2	10	16.560	16.560
(b)	4	10	33.120	33.12
(c)	6	100	496.800	496.8
(d)	5	50	207.000	207.0

Exercise 7.12

Both the direct-reading and null-balance strain indicators incorporate Wheatstone bridges powered with low-voltage supplies. After initial adjustments for balance and calibration, the direct-reading indicator displays the strain on a DVM. As the load is varied, the strain reading is updated at a frequency of 1 or 2 readings/sec. The convenience of direct display without operator interaction is clear.

With the null-balance method, the same initial adjustments for balance and calibration is required. However, when the strain on a gage changes, a meter on the indicator shows that the bridge is out of balance. The operator then brings the bridge back into balance and reads the amount of adjustment required for balance in units of strain. The null-balance method is accurate and the instrument is less expensive than the direct-reading indicator. The disadvantage is in the operator time necessary for balance and the lack of a direct display.

Exercise 7.13

From Eqs. (5.5) and (6.18):
$$v_o = \frac{r}{(1+r)^2} v_s S_g G \varepsilon$$

For the bar: $\varepsilon_B = \frac{P}{AE} = C_1 P$ (depends on bar size and material)

$$\varepsilon = C_2 \varepsilon_B = C_1 C_2 P = KP \text{ (depends on gage placement)}$$

With $r = 1$ (4 gages), $S_g = 2.06$, and $v_s = 4$ V:
$$v_o = \frac{1}{(1+1)^2}(4)(2.06)GKP = 2.06 GKP$$

With $G = 20$, $v_o = 6280$ for $P = 10,000$ lb:
$$K = \frac{v_o}{2.06 GP} = \frac{6280}{2.06(20)(10,000)} = 0.015243$$

For $v_o = 10,000$:
$$G = \frac{v_o}{2.06 KP} = \frac{10,000}{2.06(0.015243)(10,000)} = 31.85$$

Reading will correspond to the load if G is set at 31.85.

Exercise 7.14

From Eqs. (6.18) and (5.5):
$$v_o = \frac{r}{(1+r)^2} S_g v_s \varepsilon = \frac{5}{(1+5)^2}(3.35)(9)(900)(10^{-6})$$
$$= 3.769(10^{-3}) \text{ V} = 3.769 \text{ mV}$$

$$S_R = \frac{v_o}{d_s} = \frac{3.769}{4} = 0.942 \text{ mV/div.}$$

Exercise 7.15

From Eqs. (6.18) and (5.5):
$$v_o = \frac{r}{(1+r)^2} S_g v_s \varepsilon = \frac{5}{(1+5)^2}(3.35)(9)(1400)(10^{-6})$$
$$= 5.863(10^{-3}) \text{ V} = 5.863 \text{ mV}$$

$$d_s = \frac{v_o}{S_R} = \frac{5.863}{1} = 5.86 \text{ div.}$$

Exercise 7.16

$$v_o = S_R d_s = 1(5.2) = 5.2 \text{ mV}$$

From Eqs. (6.18) and (5.5):
$$v_o = \frac{r}{(1 + r)^2} S_g v_s \varepsilon$$

$$\varepsilon = \frac{v_o(1 + r)^2}{r S_g v_s} = \frac{5.2(10^{-3})(1 + 5)^2}{5(3.35)(9)} = 1242(10^{-6}) \text{ m/m} = 1242 \text{ } \mu\text{m/m}$$

Exercise 7.17

From Eq. (7.12):
$$R_B = \frac{r}{1 + r}\left(\frac{q + 1}{q}\right) R_g$$

With $R_1 = R_2 = R_3 = R_4 = R_g$: $\quad r = R_2/R_1 = 1 \quad$ and $\quad q = R_2/R_3 = 1$

$$R_B = \frac{1}{1 + 1}\left(\frac{1 + 1}{1}\right) R_g = R_g$$

From Eq. (7.18):
$$S_s = \frac{1}{2} \sqrt{p_g/R_g} \; S_g \frac{R_g}{R_g + R_G} S_G$$

$$S_s = \frac{1}{2} \sqrt{0.25/350} \; (2.07) \frac{350}{350 + 100} (0.003)$$

$$= 64.5 \text{ mm/(m/m)} = 64.5(10^{-6}) \text{ mm/}(\mu\text{m/m})$$

Exercise 7.18

From Eq. (7.18):
$$S_s = \frac{1}{2} \sqrt{p_g/R_g} \; S_g \frac{R_g}{R_g + R_G} S_G$$

$$= \frac{1}{2} \sqrt{0.25/500} \; (2.07) \frac{500}{500 + 100} (0.003)$$

$$= 57.9 \text{ mm/(m/m)} = 57.9(10^{-6}) \text{ mm/}(\mu\text{m/m})$$

The sensitivity is not improved; it decreases.

$$\% \text{ decrease} = \frac{64.5 - 57.9}{64.5} (100) = 10.2\%$$

Provided the galvanometer requires an external resistance of 500 Ω.

Exercise 7.19

From Eq. (7.21): $\quad \varepsilon = Cd_s = 40(26) = 1040 \; \mu m/m$

Exercise 7.20

From Eq. (7.20):

$$C = \frac{(1+r)^2 S_A S_R}{r v_s S_g} = \frac{(1+2)^2(10)(10^{-3})}{2(4)(2.04)(50)}$$

$$= 110.3(10^{-6}) \; (m/m)/div = 110.3 \; (\mu m/m)/div$$

Exercise 7.21

From Eq. (7.26): $\quad \varepsilon_c = \dfrac{R_2}{S_g(R_2 + R_c)}$

which yields: $\quad R_c = \dfrac{R_2}{S_g \varepsilon_c} - R_2$

or since $R_2 = rR_1 = rR_g$: $\quad R_c = \dfrac{rR_g}{S_g \varepsilon_c} - rR_g$

(a) For $\varepsilon_c = 600 \; \mu m/m$, $r = 3$, $R_g = 350 \; \Omega$, and $S_g = 2.06$:

$$R_c = \frac{3(350)}{2.06(600)(10^{-6})} - 3(350) = 848(10^3) \; \Omega = 848 \; k\Omega$$

(b) For $\varepsilon_c = 1000 \; \mu m/m$, $r = 2$, $R_g = 500 \; \Omega$, and $S_g = 2.07$:

$$R_c = \frac{2(500)}{2.07(1000)(10^{-6})} - 2(500) = 482(10^3) \; \Omega = 482 \; k\Omega$$

(c) For $\varepsilon_c = 900 \; \mu m/m$, $r = 1$, $R_g = 120 \; \Omega$, and $S_g = 2.05$:

$$R_c = \frac{1(120)}{2.05(900)(10^{-6})} - 1(120) = 64.9(10^3) \; \Omega = 64.9 \; k\Omega$$

(d) For $\varepsilon_c = 2000 \; \mu m/m$, $r = 2$, $R_g = 350 \; \Omega$, and $S_g = 2.09$:

$$R_c = \frac{2(350)}{2.09(2000)(10^{-6})} - 2(350) = 166.8(10^3) \; \Omega = 166.8 \; k\Omega$$

Exercise 7.22

$$\sigma = \frac{Mc}{I} = \frac{Pah}{2I}$$

$$\varepsilon = \frac{\sigma}{E} = \frac{Pah}{2EI}$$

$$\delta = \frac{PL^3}{3EI} = \frac{2L^3\varepsilon}{3ah}$$

$$\varepsilon = \frac{3ah\delta}{2L^3}$$

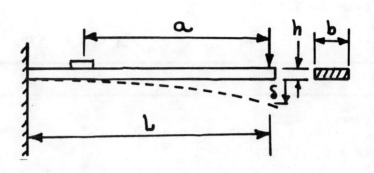

With an aluminum beam: $E = 10\,(10^6)$ psi

$$\varepsilon_{yield} = 3000 \; \mu in./in.$$

With $L = 12$ in., $a = 10$ in., $h = 0.25$ in., and $b = 1.00$ in.

$$\varepsilon = \frac{3ah\delta}{2L^3} = \frac{3(10)(0.25)}{2(12)^3}\,\delta = 2.17(10^{-3})\,\delta$$

From which: $\delta = 460.8\,\varepsilon$

With:
$\varepsilon = 500\;\mu in./in.$ $\delta = 0.230$ in.
$\varepsilon = 1000\;\mu in./in.$ $\delta = 0.461$ in.
$\varepsilon = 1500\;\mu in./in.$ $\delta = 0.691$ in.
$\varepsilon = 2000\;\mu in./in.$ $\delta = 0.922$ in.

With $a = 9.216$ in.:
$\varepsilon = 500\;\mu in./in.$ $\delta = 0.250$ in.
$\varepsilon = 1000\;\mu in./in.$ $\delta = 0.500$ in.
$\varepsilon = 1500\;\mu in./in.$ $\delta = 0.750$ in.
$\varepsilon = 2000\;\mu in./in.$ $\delta = 1.000$ in.

Standard size gage blocks can be used to control δ.

Exercise 7.23

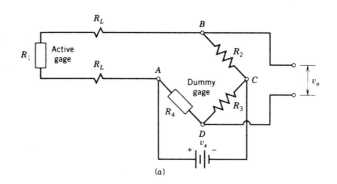

From Eq. (7.27):
$$\mathcal{L} = \frac{2R_L/R_g}{1 + (2R_L/R_g)}$$

(a) From Table 7.2:

$R_L = 10.310(1.90) = 19.589 \; \Omega$

$2R_L/R_g = 2(19.589)/120 = 0.3265$

$\mathcal{L} = \frac{0.3265}{1 + 0.3265}(100) = 24.6\%$

(b) From Table 7.2:

$R_L = 4.081(1.90) = 7.754 \; \Omega$

$2R_L/R_g = 2(7.754)/120 = 0.1292$

$\mathcal{L} = \frac{0.1292}{1 + 0.1292}(100) = 11.4\%$

(c) From Table 7.2:

$R_L = 1.015(1.90) = 1.9285 \; \Omega$

$2R_L/R_g = 2(1.9285)/120 = 0.03214$

$\mathcal{L} = \frac{0.03214}{1 + 0.03214}(100) = 3.11\%$

(d) From Table 7.2:

$R_L = 0.253(1.90) = 0.4807 \; \Omega$

$2R_L/R_g = 2(0.4807)/120 = 0.00801$

$\mathcal{L} = \frac{0.00801}{1 + 0.00801}(100) = 0.795\%$

Exercise 7.24

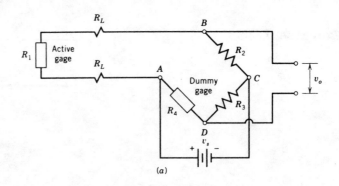

From Eq. (7.27):
$$\mathcal{L} = \frac{2R_L/R_g}{1 + (2R_L/R_g)}$$

(a) From Table 7.2:
$$R_L = 10.310(1.90) = 19.589 \ \Omega$$
$$2R_L/R_g = 2(19.589)/350 = 0.1119$$
$$\mathcal{L} = \frac{0.1119}{1 + 0.1119}(100) = 10.1\%$$

(b) From Table 7.2:
$$R_L = 4.081(1.90) = 7.754 \ \Omega$$
$$2R_L/R_g = 2(7.754)/350 = 0.0443$$
$$\mathcal{L} = \frac{0.0443}{1 + 0.0443}(100) = 4.24\%$$

(c) From Table 7.2:
$$R_L = 1.015(1.90) = 1.9285 \ \Omega$$
$$2R_L/R_g = 2(1.9285)/350 = 0.01102$$
$$\mathcal{L} = \frac{0.01102}{1 + 0.01102}(100) = 1.09\%$$

(d) From Table 7.2:
$$R_L = 0.253(1.90) = 0.4807 \ \Omega$$
$$2R_L/R_g = 2(0.4807)/350 = 0.00275$$
$$\mathcal{L} = \frac{0.00275}{1 + 0.00275}(100) = 0.274\%$$

Exercise 7.25

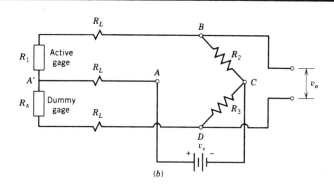

From Eq. (7.29):
$$\mathcal{L} = \frac{R_L/R_g}{1 + (R_L/R_g)}$$

(a) From Table 7.2:

$R_L = 10.310(1.90) = 19.589 \ \Omega$

$R_L/R_g = 19.589/120 = 0.1632$

$\mathcal{L} = \frac{0.1632}{1 + 0.1632}(100) = 14.0\%$

(b) From Table 7.2:

$R_L = 4.081(1.90) = 7.754 \ \Omega$

$R_L/R_g = 7.754/120 = 0.0646$

$\mathcal{L} = \frac{0.0646}{1 + 0.0646}(100) = 6.07\%$

(c) From Table 7.2:

$R_L = 1.015(1.90) = 1.9285 \ \Omega$

$R_L/R_g = 1.9285/120 = 0.01607$

$\mathcal{L} = \frac{0.01607}{1 + 0.01607}(100) = 1.58\%$

(d) From Table 7.2:

$R_L = 0.253(1.90) = 0.4807 \ \Omega$

$R_L/R_g = 0.4807/120 = 0.00401$

$\mathcal{L} = \frac{0.00401}{1 + 0.00401}(100) = 0.399\%$

Exercise 7.26

From Eq. (7.31):
$$\varepsilon' = \frac{\Delta R_s / R_g}{S_g}$$

With $R_g = 120 \, \Omega$ and $S_g = 2.09$:
$$\varepsilon' = \frac{\Delta R_s}{120(2.09)} = 0.003987 \Delta R_s$$

ΔR_s ($\mu\Omega$)	ε' (μin./in.)
100	0.4
1000	4.0
10,000	39.9
100,000	398.7

Exercise 7.27

When strain gages are employed on rotating elements which have a shaft with one end free, slip rings may be used in the connections between the shaft and the fixed instrument station. The slip ring assembly has two main components:

1. An inner core consisting of a series of rings each electrically insulated from the other.
2. An outer shell that contains the brushes that make electrical contact with the rings.

The inner core is fastened to the shaft end and it rotates with the shaft. The outer shell, which does not rotate, has lead wires that connect with the instruments (power supply and bridge). The lead wires from the gages connect to the slip rings on the inner core. The electrical interface between the rotating shaft and the stationary instruments is made at the sliding contact between the brushes and rings. This sliding contact is not perfect and the electrical resistance between the brush and the ring fluctuates. While several brushes (in parallel) are used against each ring to mitigate these fluctuations, ΔR is still excessive. For this reason a complete bridge is placed on the rotating member for each active gage and the slip rings are used only to connect the external leads (two output and two input) of the bridge.

Exercise 7.28

The specification should contain the following paragraphs:

1. Why electrical noise is a problem.
2. Sources of noise - avoiding ground loops.
3. Precautions:
   a. twisted leads.
   b. shielded leads.
   c. differential amplifiers.
4. Measurements to make to test the system.
5. Determining signal to noise ratio.

Exercise 7.29

Twisted leads, shielded leads with instrument end grounding, and differential amplifiers.

Exercise 7.30

$$\varepsilon_T = (6 - 13)(10^{-6})(30) = -210(10^{-6}) \text{ in./in.} = -210 \, \mu\text{in./in.}$$

$$\varepsilon_{total} = \varepsilon_\sigma + \varepsilon_T = 200 \, \mu\text{in./in.}$$

$$\varepsilon_\sigma = 200 - (-210) = 410 \, \mu\text{in./in.}$$

$$\%\varepsilon = \frac{410 - 200}{410}(100) = 51.2\%$$

Exercise 7.31

From Eq. (7.38):

$$S_a = \frac{S_g}{1 - \nu_0 K_t} = \frac{2.04}{1 - 0.285(0.02)} = 2.05$$

Exercise 7.32

For a thin-walled cylindrical pressure vessel:

$$\sigma_a = \frac{pr}{2t} \qquad \sigma_h = \frac{pr}{t} \qquad \sigma_h = 2\sigma_a$$

$$\varepsilon_a = \frac{1}{E}\left[\sigma_a - \nu(2\sigma_a)\right] = \frac{\sigma_a}{E}(1 - 2\nu)$$

$$\varepsilon_h = \frac{1}{E}\left[2\sigma_a - \nu\sigma_a\right] = \frac{\sigma_a}{E}(2 - \nu)$$

Therefore:
$$\frac{\varepsilon_a}{\varepsilon_h} = \frac{1 - 2\nu}{2 - \nu}$$

For steel $\nu = 0.29$:

$$\frac{\varepsilon_a}{\varepsilon_h} = \frac{1 - 2\nu}{2 - \nu} = \frac{1 - 0.58}{2 - 0.29} = 0.2456$$

From Eq. (7.42):

$$\mathscr{E} = \frac{K_t(\varepsilon_a/\varepsilon_h + \nu_0)}{1 - \nu_0 K_t}(100) = \frac{0.03(0.2456 + 0.285)}{1 - 0.285(0.03)}(100) = 1.606\%$$

Exercise 7.33

From Eqs. (7.46):

$$\varepsilon_{xx} = \frac{1 - \nu_0 K_t}{1 - K_t^2}\left[\varepsilon'_{xx} - K_t \varepsilon'_{yy}\right]$$

$$\varepsilon_{yy} = \frac{1 - \nu_0 K_t}{1 - K_t^2}\left[\varepsilon'_{yy} - K_t \varepsilon'_{xx}\right]$$

For gages with $K_t = 0.04$:

$$\varepsilon_{xx} = \frac{1 - 0.285(0.04)}{1 - (0.04)^2}\left[\varepsilon'_{xx} - 0.04\varepsilon'_{yy}\right] = 0.990\left[\varepsilon'_{xx} - 0.04\varepsilon'_{yy}\right]$$

$$\varepsilon_{yy} = \frac{1 - 0.285(0.04)}{1 - (0.04)^2}\left[\varepsilon'_{yy} - 0.04\varepsilon'_{xx}\right] = 0.990\left[\varepsilon'_{yy} - 0.04\varepsilon'_{xx}\right]$$

Exercise 7.33 (Continued)

	Apparent Strain		True Strain	
	ε'_{xx}	ε'_{yy}	ε_{xx}	ε_{yy}
	µin./in.	µin./in.	µin./in.	µin./in.
1	800	1200	744	1156
2	640	-720	662	-738
3	1120	-240	1118	-282
4	-560	2400	-649	2398
5	240	1440	181	1416

Exercise 7.34

For a simple tension test: $\qquad \varepsilon_t = -\nu \varepsilon_a$

From Eq. (7.45):

$$\varepsilon'_a = \frac{1}{1 - \nu_o K_t}\left[\varepsilon_a - \nu K_t \varepsilon_a\right] \qquad \varepsilon'_t = \frac{1}{1 - \nu_o K_t}\left[\varepsilon_t - (K_t/\nu)\varepsilon_t\right]$$

$$\varepsilon'_a = \frac{1 - \nu K_t}{1 - \nu_o K_t}\varepsilon_a \qquad \nu' = -\frac{\varepsilon'_t}{\varepsilon'_a}$$

$$\varepsilon'_t = \frac{1 - K_t/\nu}{1 - \nu_o K_t}\varepsilon_t \qquad \nu = -\frac{\varepsilon_t}{\varepsilon_a}$$

$$\nu' = \frac{1 - K_t/\nu}{1 - \nu K_t}\nu = \frac{1 - 0.03/\nu}{1 - 0.03\nu}(\nu) = \frac{\nu - 0.03}{1 - 0.03\nu}$$

$$\%\mathcal{E} = \frac{\nu' - \nu}{\nu}(100) = \frac{0.03(\nu^2 - 1)}{\nu(1 - 0.03\nu)}(100)$$

ν	Error %
0.20	-14.5
0.25	-11.3
0.30	-9.2
0.33	-8.2
0.35	-7.6
0.40	-6.4

Exercise 7.35

Consider a state of stress with principal axes u and v. Along a direction n oriented at an angle θ with respect to the u-axis:

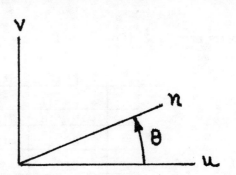

$$\sigma_n = \sigma_u \cos^2 \theta + \sigma_v \sin^2 \theta$$
$$\sigma_t = \sigma_u \sin^2 \theta + \sigma_v \cos^2 \theta$$

Thus:
$$\varepsilon_n = \frac{\sigma_n}{E} - \frac{\nu \sigma_t}{E} = \frac{\sigma_u}{E}(\cos^2 \theta - \nu \sin^2 \theta) + \frac{\sigma_v}{E}(\sin^2 \theta - \nu \cos^2 \theta)$$

When $\sin^2 \theta - \nu \cos^2 \theta = 0$: $\qquad \varepsilon_n \sim \sigma_u$

and $\qquad \tan^2 \theta = \dfrac{\sin^2 \theta}{\cos^2 \theta} = \nu \qquad\qquad \theta = \tan^{-1} \sqrt{\nu}$

Therefore: $\qquad \sin \theta = \dfrac{\sqrt{\nu}}{\sqrt{1 + \nu^2}} \qquad\qquad \cos \theta = \dfrac{1}{\sqrt{1 + \nu^2}}$

$$\varepsilon_n = \frac{\sigma_u}{E}\left(\frac{1}{1 + \nu^2} - \frac{\nu^2}{1 + \nu^2}\right) = \frac{(1 - \nu^2)\sigma_u}{(1 + \nu^2)E}$$

$$\frac{\Delta R}{R} = S_g \varepsilon_n = S_{sg} \sigma_u \qquad \text{therefore} \qquad S_{sg} = \frac{S_g(1 - \nu^2)}{E(1 + \nu^2)}$$

Exercise 7.36

A commercially available stress gage is shown in Fig. 5.9m. For such a gage, with the axis of the gage oriented as shown in the figure at the right, the strain along the top grid element is given by a modified form of eq. (7.49) as

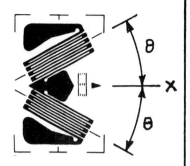

$$\varepsilon_{\phi-\theta} = \frac{1}{2}(\varepsilon_1 + \varepsilon_2) + \frac{1}{2}(\varepsilon_1 - \varepsilon_2)\cos 2(\phi - \theta) \qquad (a)$$

Similarly, the strain along the lower grid element is

$$\varepsilon_{\phi+\theta} = \frac{1}{2}(\varepsilon_1 + \varepsilon_2) + \frac{1}{2}(\varepsilon_1 - \varepsilon_2)\cos 2(\phi + \theta) \qquad (b)$$

Summing Eqs. (a) and (b) and expanding the cosine terms yield

$$\varepsilon_{\phi-\theta} + \varepsilon_{\phi+\theta} = (\varepsilon_1 + \varepsilon_2) + (\varepsilon_1 - \varepsilon_2)\cos 2\phi \cos 2\theta \qquad (c)$$

Recall that
$$\varepsilon_{xx} + \varepsilon_{yy} = \varepsilon_1 + \varepsilon_2 \qquad (d)$$

$$\varepsilon_{xx} - \varepsilon_{yy} = (\varepsilon_1 - \varepsilon_2)\cos 2\phi \qquad (e)$$

Thus:
$$\varepsilon_{\phi-\theta} + \varepsilon_{\phi+\theta} = (\varepsilon_{xx} + \varepsilon_{yy}) + (\varepsilon_{xx} - \varepsilon_{yy})\cos 2\theta$$
$$= 2(\varepsilon_{xx}\cos^2\theta + \varepsilon_{yy}\sin^2\theta)$$
$$= 2\cos^2\theta\,(\varepsilon_{xx} + \varepsilon_{yy}\tan^2\theta) \qquad (f)$$

If the gage is manufactured so that θ is equal to $\tan^{-1}\sqrt{\nu}$, then

$$\tan^2\theta = \nu \qquad \cos^2\theta = \frac{1}{1+\nu}$$

and Eq. (f) becomes
$$\varepsilon_{\phi-\theta} + \varepsilon_{\phi+\theta} = \frac{2}{1+\nu}(\varepsilon_{xx} + \nu\varepsilon_{yy}) \qquad (g)$$

For a two-dimensional state of stress,

$$\sigma_{xx} = \frac{E}{(1-\nu^2)}(\varepsilon_{xx} + \nu\varepsilon_{yy})$$

Thus,
$$\sigma_{xx} = \frac{E}{2(1-\nu)}(\varepsilon_{\phi-\theta} + \varepsilon_{\phi+\theta}) = \frac{E}{(1-\nu)}\frac{\Delta R/R}{S_g}$$

With $E = 29{,}000{,}000$ psi, $\nu = 0.29$, $\sigma_{xx} = 50{,}000$ psi, and $S_g = 2.00$:

$$\frac{\Delta R}{R} = \frac{(1-\nu)S_g\sigma_{xx}}{E} = \frac{(1-0.29)(2.00)(50{,}000)}{29{,}000{,}000} = 2448(10^{-6})$$

ENGINEERING MEASUREMENTS by J. W. DALLY, W. F. RILEY, AND K. G. McCONNELL

Exercise 7.37

From Eq. (7.47): $\sigma = E\varepsilon$

For steel (E = 29,000 ksi and ν = 0.29):

$$\sigma_1 = 29(10^6)(800)(10^{-6}) = 23,200 \text{ psi} \qquad \sigma_2 = \sigma_3 = 0$$

For aluminum (E = 10,000 ksi and ν = 0.33):

$$\sigma_1 = 10(10^6)(1100)(10^{-6}) = 11,000 \text{ psi} \qquad \sigma_2 = \sigma_3 = 0$$

For titanium (E = 14,000 ksi and ν = 0.25):

$$\sigma_1 = 14(10^6)(1620)(10^{-6}) = 22,680 \text{ psi} \qquad \sigma_2 = \sigma_3 = 0$$

Exercise 7.38

From Eq. (7.48): $\quad \sigma_1 = \dfrac{E}{1 - \nu^2}(\varepsilon_1 + \nu\varepsilon_2) \qquad \sigma_2 = \dfrac{E}{1 - \nu^2}(\varepsilon_2 + \nu\varepsilon_1)$

For ε_1 = -600 μin./in. and ε_2 = -900 μin./in. in Aluminum:

$$\sigma_1 = \frac{10(10^6)}{1 - 0.33^2}\left[-600(10^{-6}) + 0.33(-900)(10^{-6})\right] = -10,066 \text{ psi} = 10,066 \text{ psi C}$$

$$\sigma_2 = \frac{10(10^6)}{1 - 0.33^2}\left[-900(10^{-6}) + 0.33(-600)(10^{-6})\right] = -12,322 \text{ psi} = 12,322 \text{ psi C}$$

For ε_1 = 1220 μin./in. and ε_2 = -470 μin./in. in Titanium:

$$\sigma_1 = \frac{14(10^6)}{1 - 0.25^2}\left[1220(10^{-6}) + 0.25(-470)(10^{-6})\right] = 16,464 \text{ psi} = 16,464 \text{ psi T}$$

$$\sigma_2 = \frac{14(10^6)}{1 - 0.25^2}\left[-470(10^{-6}) + 0.25(1220)(10^{-6})\right] = -2464 \text{ psi} = 2464 \text{ psi C}$$

For ε_1 = 1115 μin./in. and ε_2 = 820 μin./in. in Steel:

$$\sigma_1 = \frac{29(10^6)}{1 - 0.29^2}\left[1115(10^{-6}) + 0.29(820)(10^{-6})\right] = 42,833 \text{ psi} = 42,833 \text{ psi T}$$

$$\sigma_2 = \frac{29(10^6)}{1 - 0.29^2}\left[820(10^{-6}) + 0.29(1115)(10^{-6})\right] = 36,202 \text{ psi} = 36,202 \text{ psi T}$$

Exercise 7.39

1. For
$$\varepsilon_A = \varepsilon_{xx} = 600 \; \mu\text{in./in.} \qquad E = 10,000 \text{ ksi}$$
$$\varepsilon_B = 1200 \; \mu\text{in./in.} \qquad \nu = 0.33$$
$$\varepsilon_C = \varepsilon_{yy} = -300 \; \mu\text{in./in.}$$

$$\gamma_{xy} = 2\varepsilon_B - \varepsilon_A - \varepsilon_C = 2(1200) - 600 - (-300) = 2100 \; \mu\text{in./in.}$$

From Eq. (7.50):
$$\varepsilon_{1,2} = \frac{\varepsilon_{xx} + \varepsilon_{yy}}{2} \pm \frac{1}{2} \sqrt{(\varepsilon_{xx} - \varepsilon_{yy})^2 + \gamma_{xy}^2}$$

$$= \frac{600 - 300}{2} \pm \frac{1}{2} \sqrt{(600 + 300)^2 + (2100)^2}$$

$$\varepsilon_1 = 150 + 1142 = 1292 \; \mu\text{in./in.}$$
$$\varepsilon_2 = 150 - 1142 = -992 \; \mu\text{in./in.}$$

From Eq. (7.48):

$$\sigma_1 = \frac{E}{1-\nu^2}(\varepsilon_1 + \nu\varepsilon_2) = \frac{10(10^6)}{1 - 0.33^2}\Big[1292 + 0.33(-992)\Big](10^{-6}) = 10,825 \text{ psi}$$

$$\sigma_2 = \frac{E}{1-\nu^2}(\varepsilon_2 + \nu\varepsilon_1) = \frac{10(10^6)}{1 - 0.33^2}\Big[-992 + 0.33(1292)\Big](10^{-6}) = -6,348 \text{ psi}$$

$$\phi = \frac{1}{2}\tan^{-1}\frac{\gamma_{xy}}{\varepsilon_x - \varepsilon_y} = \frac{1}{2}\tan^{-1}\frac{2100}{600 + 300} = 33.4°$$

2. For
$$\varepsilon_A = \varepsilon_{xx} = 1050 \; \mu\text{in./in.} \qquad E = 29,000 \text{ ksi}$$
$$\varepsilon_B = 1050 \; \mu\text{in./in.} \qquad \nu = 0.29$$
$$\varepsilon_C = \varepsilon_{yy} = 1050 \; \mu\text{in./in.}$$

$$\gamma_{xy} = 2\varepsilon_B - \varepsilon_A - \varepsilon_C = 2(1050) - 1050 - 1050 = 0$$

From Eq. (7.50):
$$\varepsilon_{1,2} = \frac{\varepsilon_{xx} + \varepsilon_{yy}}{2} \pm \frac{1}{2} \sqrt{(\varepsilon_{xx} - \varepsilon_{yy})^2 + \gamma_{xy}^2}$$

$$= \frac{1050 + 1050}{2} \pm \frac{1}{2} \sqrt{(1050 - 1050)^2 + (0)^2}$$

$$\varepsilon_1 = 1050 + 0 = 1050 \; \mu\text{in./in.}$$
$$\varepsilon_2 = 1050 - 0 = 1050 \; \mu\text{in./in.}$$

Exercise 7.39 (Continued)

From Eq. (7.48):

$$\sigma_1 = \frac{E}{1-\nu^2}(\epsilon_1 + \nu\epsilon_2) = \frac{29(10^6)}{1-0.29^2}\Big[1050 + 0.29(1050)\Big](10^{-6}) = 42{,}887 \text{ psi}$$

$$\sigma_2 = \frac{E}{1-\nu^2}(\epsilon_2 + \nu\epsilon_1) = \frac{29(10^6)}{1-0.29^2}\Big[1050 + 0.29(1050)\Big](10^{-6}) = 42{,}887 \text{ psi}$$

$$\phi = \tfrac{1}{2}\tan^{-1}\frac{\gamma_{xy}}{\epsilon_x - \epsilon_y} = \tfrac{1}{2}\tan^{-1}\frac{0}{1050-1050} = \text{indeterminate}$$

All planes with outer normals in the xy plane are principal.

3. For $\quad \epsilon_A = \epsilon_{xx} = -450\ \mu\text{in./in.} \qquad E = 14{,}000 \text{ ksi}$

$\qquad\qquad\qquad \epsilon_B = -900\ \mu\text{in./in.} \qquad\qquad \nu = 0.25$

$\qquad\qquad\qquad \epsilon_C = \epsilon_{yy} = 1350\ \mu\text{in./in.}$

$\gamma_{xy} = 2\epsilon_B - \epsilon_A - \epsilon_C = 2(-900) - (-450) - (1350) = -2700\ \mu\text{in./in.}$

From Eq. (7.50):

$$\epsilon_{1,2} = \frac{\epsilon_{xx} + \epsilon_{yy}}{2} \pm \tfrac{1}{2}\sqrt{(\epsilon_{xx} - \epsilon_{yy})^2 + \gamma_{xy}^2}$$

$$= \frac{-450 + 1350}{2} \pm \tfrac{1}{2}\sqrt{(-450 - 1350)^2 + (-2700)^2}$$

$\epsilon_1 = 450 + 1622 = 2072\ \mu\text{in./in.}$

$\epsilon_2 = 450 - 1622 = -1172\ \mu\text{in./in.}$

From Eq. (7.48):

$$\sigma_1 = \frac{E}{1-\nu^2}(\epsilon_1 + \nu\epsilon_2) = \frac{14(10^6)}{1-0.25^2}\Big[2072 + 0.25(-1172)\Big](10^{-6}) = 26{,}566 \text{ psi}$$

$$\sigma_2 = \frac{E}{1-\nu^2}(\epsilon_2 + \nu\epsilon_1) = \frac{14(10^6)}{1-0.25^2}\Big[-1172 + 0.25(2072)\Big](10^{-6}) = -9{,}766 \text{ psi}$$

$$\phi = \tfrac{1}{2}\tan^{-1}\frac{\gamma_{xy}}{\epsilon_x - \epsilon_y} = \tfrac{1}{2}\tan^{-1}\frac{-2700}{-450-1350} = 28.2°$$

Exercise 7.40

From Eqs. (7.49):

$$\varepsilon_A = \varepsilon_{xx}\cos^2\theta_A + \varepsilon_{yy}\sin^2\theta_A + \gamma_{xy}\sin\theta_A\cos\theta_A$$

$$\varepsilon_B = \varepsilon_{xx}\cos^2\theta_B + \varepsilon_{yy}\sin^2\theta_B + \gamma_{xy}\sin\theta_B\cos\theta_B$$

$$\varepsilon_C = \varepsilon_{xx}\cos^2\theta_C + \varepsilon_{yy}\sin^2\theta_C + \gamma_{xy}\sin\theta_C\cos\theta_C$$

For the three-element rectangular rosette shown at the right:

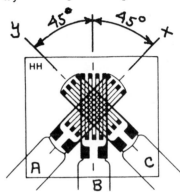

$\theta_A = 0°$ $\quad \varepsilon_A = \varepsilon_{xx}$

$\theta_B = 45°$ $\quad \varepsilon_B = \frac{1}{2}(\varepsilon_{xx} + \varepsilon_{yy} + \gamma_{xy})$

$\theta_C = 90°$ $\quad \varepsilon_C = \varepsilon_{yy}$

From which:
$$\gamma_{xy} = 2\varepsilon_B - \varepsilon_A - \varepsilon_C$$

From the strain transformation equations (Eqs. 7.50) for a two-dimensional state of stress:

$$\varepsilon_1 = \frac{1}{2}(\varepsilon_{xx} + \varepsilon_{yy}) + \frac{1}{2}\sqrt{(\varepsilon_{xx} - \varepsilon_{yy})^2 + \gamma_{xy}^2}$$

$$\varepsilon_2 = \frac{1}{2}(\varepsilon_{xx} + \varepsilon_{yy}) - \frac{1}{2}\sqrt{(\varepsilon_{xx} - \varepsilon_{yy})^2 + \gamma_{xy}^2}$$

$$\phi = \frac{1}{2}\tan^{-1}\frac{\gamma_{xy}}{\varepsilon_{xx} - \varepsilon_{yy}}$$

Thus:

$$\varepsilon_1 = \frac{1}{2}(\varepsilon_A + \varepsilon_C) + \frac{1}{2}\sqrt{(\varepsilon_A - \varepsilon_C)^2 + (2\varepsilon_B - \varepsilon_A - \varepsilon_C)^2}$$

$$\varepsilon_2 = \frac{1}{2}(\varepsilon_A + \varepsilon_C) - \frac{1}{2}\sqrt{(\varepsilon_A - \varepsilon_C)^2 + (2\varepsilon_B - \varepsilon_A - \varepsilon_C)^2}$$

$$\phi = \frac{1}{2}\tan^{-1}\frac{2\varepsilon_B - \varepsilon_A - \varepsilon_C}{\varepsilon_A - \varepsilon_C}$$

Exercise 7.41

From Eqs. (7.49):

$$\varepsilon_A = \varepsilon_{xx}\cos^2\theta_A + \varepsilon_{yy}\sin^2\theta_A + \gamma_{xy}\sin\theta_A\cos\theta_A$$

$$\varepsilon_B = \varepsilon_{xx}\cos^2\theta_B + \varepsilon_{yy}\sin^2\theta_B + \gamma_{xy}\sin\theta_B\cos\theta_B$$

$$\varepsilon_C = \varepsilon_{xx}\cos^2\theta_C + \varepsilon_{yy}\sin^2\theta_C + \gamma_{xy}\sin\theta_C\cos\theta_C$$

For the three-element delta rosette shown at the right:

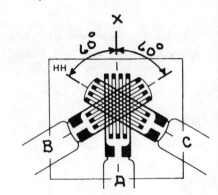

$\theta_A = 0°$ $\varepsilon_A = \varepsilon_{xx}$

$\theta_B = 120°$ $\varepsilon_B = \frac{1}{4}(\varepsilon_{xx} + 3\varepsilon_{yy} - \sqrt{3}\gamma_{xy})$

$\theta_C = 240°$ $\varepsilon_C = \frac{1}{4}(\varepsilon_{xx} + 3\varepsilon_{yy} + \sqrt{3}\gamma_{xy})$

From which:

$$\varepsilon_{yy} = \frac{1}{3}[2(\varepsilon_B + \varepsilon_C) - \varepsilon_A]$$

$$\gamma_{xy} = \frac{2\sqrt{3}}{3}(\varepsilon_C - \varepsilon_B)$$

From the strain transformation equations (Eqs. 7.50) for a two-dimensional state of stress:

$$\varepsilon_1 = \frac{1}{2}(\varepsilon_{xx} + \varepsilon_{yy}) + \frac{1}{2}\sqrt{(\varepsilon_{xx} - \varepsilon_{yy})^2 + \gamma_{xy}^2}$$

$$\varepsilon_2 = \frac{1}{2}(\varepsilon_{xx} + \varepsilon_{yy}) - \frac{1}{2}\sqrt{(\varepsilon_{xx} - \varepsilon_{yy})^2 + \gamma_{xy}^2}$$

$$\phi = \frac{1}{2}\tan^{-1}\frac{\gamma_{xy}}{\varepsilon_{xx} - \varepsilon_{yy}}$$

Thus:

$$\varepsilon_1 = \frac{1}{3}(\varepsilon_A + \varepsilon_B + \varepsilon_C) + \frac{\sqrt{2}}{3}\sqrt{(\varepsilon_A - \varepsilon_B)^2 + (\varepsilon_B - \varepsilon_C)^2 + (\varepsilon_C - \varepsilon_A)^2}$$

$$\varepsilon_2 = \frac{1}{3}(\varepsilon_A + \varepsilon_B + \varepsilon_C) - \frac{\sqrt{2}}{3}\sqrt{(\varepsilon_A - \varepsilon_B)^2 + (\varepsilon_B - \varepsilon_C)^2 + (\varepsilon_C - \varepsilon_A)^2}$$

$$\phi = \frac{1}{2}\tan^{-1}\frac{\sqrt{3}(\varepsilon_C - \varepsilon_B)}{2\varepsilon_A - (\varepsilon_B + \varepsilon_C)}$$

Exercise 7.42

Mohr's strain circle for the three-element delta rosette is shown below. The two principal angles ϕ_1 and ϕ_2 given by the equation

$$\phi = \frac{1}{2} \tan^{-1} \frac{\sqrt{3}(\varepsilon_C - \varepsilon_B)}{2\varepsilon_A - (\varepsilon_B + \varepsilon_C)}$$

are shown on the circle.

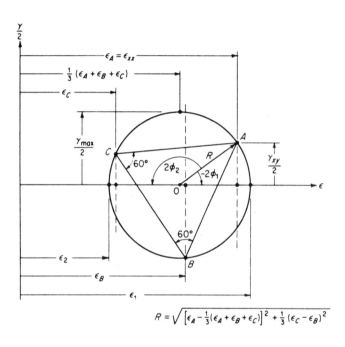

(a) From Mohr's circle it is seen that the principal angles are identified by applying the following rules:

$$\begin{array}{ll} \text{When } \varepsilon_C > \varepsilon_B & 0° < \phi_1 < 90° \\ \text{When } \varepsilon_C < \varepsilon_B & -90° < \phi_1 < 0° \\ \text{When } \varepsilon_C = \varepsilon_B \text{ and } \varepsilon_A > \varepsilon_B & \phi_1 = 0 \\ \text{When } \varepsilon_C = \varepsilon_B \text{ and } \varepsilon_A < \varepsilon_B & \phi_1 = \pm 90° \end{array}$$

(b) From Exercise 7.41:

$$\varepsilon_1 = \frac{1}{3}(\varepsilon_A + \varepsilon_B + \varepsilon_C) + \frac{\sqrt{2}}{3}\sqrt{(\varepsilon_A - \varepsilon_B)^2 + (\varepsilon_B - \varepsilon_C)^2 + (\varepsilon_C - \varepsilon_A)^2}$$

$$\varepsilon_2 = \frac{1}{3}(\varepsilon_A + \varepsilon_B + \varepsilon_C) - \frac{\sqrt{2}}{3}\sqrt{(\varepsilon_A - \varepsilon_B)^2 + (\varepsilon_B - \varepsilon_C)^2 + (\varepsilon_C - \varepsilon_A)^2}$$

Exercise 7.42 Continued)

From Eqs. (7.48):

$$\sigma_1 = \frac{E}{1-\nu^2}(\varepsilon_1 + \nu\varepsilon_2) \qquad \sigma_2 = \frac{E}{1-\nu^2}(\varepsilon_2 + \nu\varepsilon_1)$$

Thus:

$$\sigma_1 = E\left[\frac{\varepsilon_A + \varepsilon_B + \varepsilon_C}{3(1-\nu)} + \frac{\sqrt{2}}{3(1+\nu)}\sqrt{(\varepsilon_A - \varepsilon_B)^2 + (\varepsilon_B - \varepsilon_C)^2 + (\varepsilon_C - \varepsilon_A)^2}\right]$$

$$\sigma_2 = E\left[\frac{\varepsilon_A + \varepsilon_B + \varepsilon_C}{3(1-\nu)} - \frac{\sqrt{2}}{3(1+\nu)}\sqrt{(\varepsilon_A - \varepsilon_B)^2 + (\varepsilon_B - \varepsilon_C)^2 + (\varepsilon_C - \varepsilon_A)^2}\right]$$

Exercise 7.43

Mohr's strain circle for the three-element rectangular rosette is shown below. The two principal angles ϕ_1 and ϕ_2 given by the equation

$$\phi = \frac{1}{2} \tan^{-1} \frac{2\epsilon_B - \epsilon_A - \epsilon_C}{\epsilon_A - \epsilon_C}$$

are shown on the circle.

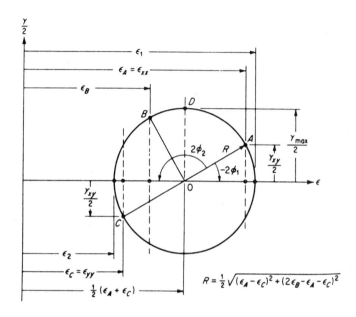

From Mohr's circle it is seen that the principal angles are identified by applying the following rules:

When $\epsilon_B > \dfrac{\epsilon_A + \epsilon_C}{2}$: $\qquad 0° < \phi_1 < 90°$

When $\epsilon_B < \dfrac{\epsilon_A + \epsilon_C}{2}$: $\qquad -90° < \phi_1 < 0°$

When $\epsilon_A > \epsilon_C$ and $\epsilon_A = \epsilon_1$: $\qquad \phi_1 = 0°$

When $\epsilon_A < \epsilon_C$ and $\epsilon_A = \epsilon_2$: $\qquad \phi_1 = \pm 90°$

Exercise 7.44

As the test temperature increases it becomes more difficult to employ electrical resistance strain gages. The difficulties are due to:

1. Changes in output with temperature T because the compensation for ΔT is limited to a small range.

2. Gage sensitivity with time at elevated temperature which causes gages to drift.

3. Degradation of the insulating characteristics of ceramic cements at very high temperatures.

Static strain measurements at temperatures up to 500°F can be made with Karma alloy gages. Static measurements at temperatures exceeding 500°F are possible but corrections must be made to account for drift by using strain-time calibrations. At these higher temperatures polymeric carriers and adhesives fail and it is necessary to use ceramic cements that are cured at temperatures in excess of the test temperature.

Dynamic strain measurements at elevated temperatures (approaching 2000° F) are possible because signal drift with time cannot occur. Again ceramic cements for bond and insulation are essential. Also, metallurgically stable alloys (no phase changes) such as Armour D or Alloy 479 are used in fabricating the gage grids.

ENGINEERING MEASUREMENTS by J. W. DALLY, W. F. RILEY, AND K. G. McCONNELL

Exercise 8.1

Transducers to measure force, torque, or pressure usually incorporate an elastic member that deforms under the imposed force, torque, or pressure. This deformation, in terms of either displacement or strain, acts as a sensor that produces an output signal. This output signal is related to the input through the properties of the elastic member and the characteristics of the sensor.

In some cases when piezoelectric or piezoresistive sensors are used, these sensors can also act as the elastic member to produce smaller more effective transducers.

The characteristics important in specifying a transducer includes its range, linearity, and sensitivity.

Exercise 8.2

Transducer	Sensor
Load	Strain gages, LVDT, piezoelectric, or piezoresistive.
Torque	Strain gages.
Pressure	Strain gages, LVDT, potentiometer, piezoelectric, or piezoresistive.

Exercise 8.3

With $E = 29,000,000$ psi,

$\nu = 0.29$,

$S_g = 2$

$v_s = 10$ V:

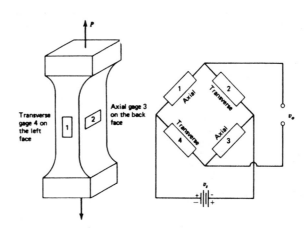

From Eq. (8.4):

$$S = \frac{S_g(1 + \nu)v_s}{2AE} = \frac{2(1 + 0.29)(10)}{2(A)(29,000,000)} = \frac{0.445(10^{-6})}{A}$$

Exercise 8.3 (Continued)

A (in.²)	S (μV/lb)
0.02	22.241
0.05	8.897
0.10	4.448
0.50	0.890
1.00	0.445
5.00	0.089
10.00	0.044
20.00	0.022

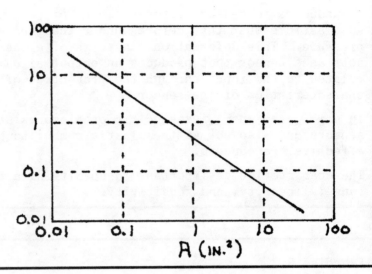

Exercise 8.4

$$p_g = v_g i_g = (v_s/2)(v_s/2R_g)$$

Therefore: $v_s = 2\sqrt{p_g R_g} = 2\sqrt{0.5(350)} = 26.5$ V

From Eq. (8.4):

$$S = \frac{S_g(1 + \nu)v_s}{2AE} = \frac{2(1 + 0.29)(26.5)}{2(A)(29,000,000)} = \frac{1.179(10^{-6})}{A}$$

A (in.²)	S (μV/lb)
0.02	58.939
0.05	23.576
0.10	11.790
0.50	2.358
1.00	1.179
5.00	0.236
10.00	0.118
20.00	0.059

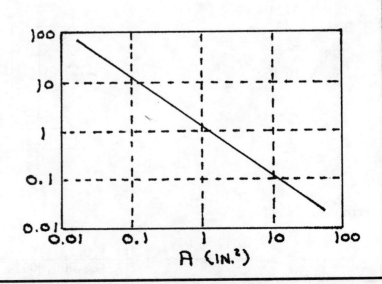

Exercise 8.5

From Eq. (8.6):

$$\frac{v_o}{v_s} = \frac{S_g S_f (1 + \nu)}{2E} = \frac{2(90,000)(1 + 0.29)}{2(29,000,000)}$$

$$= 4.00(10^{-3}) \text{ V/V} = 4.00 \text{ mV/V}$$

Exercise 8.6

For 4340 Steel: $\sigma_{yield} = 130,000$ psi

For static applications: $P_{max} = \sigma_{yield} A$

A (in.^{2})	P_{max} (lb)	A (in.^{2})	P_{max} (lb)
0.02	2600	1.00	130,000
0.05	6500	5.00	650,000
0.10	13,000	10.00	1,300,000
0.50	65,000	20.00	2,600,000

From Eq. (8.6):

$$\frac{v_o}{v_s} = \frac{S_g S_y (1 + \nu)}{2E} = \frac{2(130,000)(1 + 0.29)}{2(29,000,000)} = 5.78(10^{-3}) \text{ V/V} = 5.78 \text{ mV/V}$$

Exercise 8.7

From Eqs. (8.4) and (8.7):

$$S = \frac{v_o}{P} = \frac{v_s}{P_{max}} \left(\frac{v_o}{v_s}\right)^* = \frac{10}{100,000}(2)(10^{-3})$$

$$= 0.20(10^{-6}) \text{ V/lb} = 0.20 \text{ }\mu\text{V/lb}$$

Exercise 8.8

From Eq. (8.12):

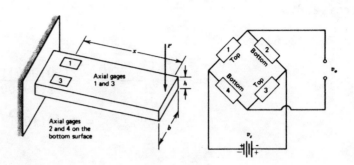

$$\left(\frac{v_o}{v_s}\right)_{max} = \frac{S_g S_f}{E} = \frac{2.00(20,000)}{10,000,000} = 4.00(10^{-3}) \text{ V/V} = 4.00 \text{mV/V}$$

From Eq. (8.10) for the beam shown in Fig. 8.3 of the text:

For the 1000-lb range:

$$\left(\frac{v_o}{v_s}\right)^* = \frac{6 S_g P_{max} x}{E b h^2} = \frac{6(2.00)(1000)x}{10(10^6) b h^2} = 0.002$$

From which $\quad x/bh^2 = 1.6667$

Practical dimensions are: $\quad x = 4h \quad$ and $\quad b = 2h$

Thus, $\quad \dfrac{x}{bh^2} = \dfrac{4h}{2h(h^2)} = 1.6667 \quad\quad h = 1.095$ in.

$$x = 4h = 4(1.095) = 4.380 \text{ in.}$$
$$b = 2h = 2(1.095) = 2.190 \text{ in.}$$
$$bh^2 = 2.190(1.095)^2 = 2.626 \text{ in}^3.$$

For the 500-lb load:

$$\left(\frac{v_o}{v_s}\right)^* = \frac{6(2.00)(500)x}{10(10^6)(2.626)} = 0.003 \quad\quad x = 13.13 \text{ in.}$$

For the 200-lb load:

$$\left(\frac{v_o}{v_s}\right)^* = \frac{6(2.00)(200)x}{10(10^6)(2.626)} = 0.004 \quad\quad x = 43.8 \text{ in.}$$

Exercise 8.9

From Eq. (8.13):

$$\delta = \frac{1.79(P)(R/t)^3}{Ew} = \frac{1.79(10,000)(10)^3}{30,000,000\, w} = \frac{0.5967}{w}$$

With $w = 3.00$ in. and $t = 1.000$ in.:

$$\delta_{max} = \frac{0.5967}{w} = \frac{0.5967}{3.00} = 0.1989 \text{ in.} \cong 0.200 \text{ in.}$$

From Eq. (8.18):

$$\sigma_\theta = 1.09\,\frac{P(R/t)}{wt} = \frac{1.09(10,000)(10)}{3.00(1.000)} \cong 36,300 \text{ psi}$$

With $R/t = 10$: $R = 10t = 10(1.000) = 10$ in.

Use a 200 HR LVDT (Table 5-1) with a nominal linear range of ±0.200 in., a sensitivity S of 2.5 (V/V)/in., and an input voltage v_s of 10 V.

From Eq. (8.17):

$$S_t = 1.79\,\frac{S(R/t)^3 v_s}{Ew} = \frac{1.79(2.5)(10^3)(10)}{30,000,000(3.00)}$$
$$= 0.497(10^{-3}) \text{ V/lb} = 0.497 \text{ mV/lb}$$

Exercise 8.10

From Exercise 8.9: $R = 10$ in. $t = 1.000$ in.
                      $w = 3.00$ in. $S = 2.5$ (V/V)/in.

From Eq. (8.19) with $S_f = 85,000$ psi:

$$P_{max} = 0.92\,\frac{wt^2 S_f}{R} = \frac{0.92(3.00)(1.000)^2(85,000)}{10} = 23,460 \text{ lb}$$

From Eq. (8.15):

$$\frac{(v_o/v_s)}{P} = \frac{1.79 S(R/t)^3}{Ew} = \frac{1.79(2.5)(10^3)}{30,000,000(3.00)}$$
$$= 49.7(10^{-6}) \text{ (V/V)/lb} = 49.7 \text{ (}\mu\text{V/V)/lb}$$

From Eq. (8.7):

$$\left(\frac{v_o}{E_s}\right)^* = \frac{(v_o/v_s)}{P} P_{max} = 49.7(10^{-6})(23,460) = 1.166 \text{ V/V}$$

Exercise 8.11

For axial loads:
$$\varepsilon_a = \frac{P}{AE} \qquad \varepsilon_t = -\nu\varepsilon_a = -\frac{\nu P}{AE}$$

$$\varepsilon_{g1} = \varepsilon_{g2} = \varepsilon_{g3} = \varepsilon_{g4} = \varepsilon_g = \varepsilon_a \cos^2 45° + \varepsilon_t \sin^2 45° = \frac{P(1-\nu)}{2AE}$$

From Eq. (6.18):
$$\frac{\Delta v_o}{v_s} = \frac{1}{4}\left[\frac{\Delta R_1}{R_1} - \frac{\Delta R_2}{R_2} + \frac{\Delta R_3}{R_3} - \frac{\Delta R_4}{R_4}\right]$$

Since
$$\frac{\Delta R_1}{R_1} = \frac{\Delta R_2}{R_2} = \frac{\Delta R_3}{R_3} = \frac{\Delta R_4}{R_4} = S_g \varepsilon_g \qquad \frac{\Delta v_o}{v_s} = 0$$

For bending loads:

Any moment M can be resolved into components M_x and M_y at the gage locations:

$$\sigma_{M_x} = 0 \text{ (on the neutral axis)} \qquad \sigma_{M_y} = \frac{M_y(d/2)}{\pi d^4/64} = \frac{32 M_y}{\pi d^3}$$

Therefore,
$$\varepsilon_a = \frac{32 M_y}{\pi d^3 E} \qquad \varepsilon_t = -\nu\varepsilon_a = -\frac{32\nu M_y}{\pi d^3 E}$$

$$\varepsilon_{g1} = \varepsilon_{g2} = -\varepsilon_{g3} = -\varepsilon_{g4} = \varepsilon_g = \varepsilon_a \cos^2 45° + \varepsilon_t \sin^2 45° = \frac{16 M_y(1-\nu)}{\pi d^3 E}$$

From Eq. (6.18):
$$\frac{\Delta v_o}{v_s} = \frac{1}{4}\left[\frac{\Delta R_1}{R_1} - \frac{\Delta R_2}{R_2} + \frac{\Delta R_3}{R_3} - \frac{\Delta R_4}{R_4}\right]$$

Since
$$\frac{\Delta R_1}{R_1} = \frac{\Delta R_2}{R_2} = -\frac{\Delta R_3}{R_3} = -\frac{\Delta R_4}{R_4} = S_g \varepsilon_g \qquad \frac{\Delta v_o}{v_s} = 0$$

Exercise 8.12

From Eq. (8.27):

$$S = \frac{16(1+\nu)S_g v_s}{\pi D^3 E} = \frac{16(1+0.30)(2.00)(8)}{\pi(1)^3(30{,}000{,}000)}$$

$$= 3.53(10^{-6}) \text{ V/(in.-lb)} = 3.53 \text{ }\mu\text{V/(in-lb)}$$

Exercise 8.13

$$p_g = v_g i_g = (v_s/2)(v_s/2R_g)$$

$$v_s = 2\sqrt{p_g R_g} = 2\sqrt{0.5(350)} = 26.5 \text{ V}$$

From Eq. (8.27):

$$S = \frac{16(1 + \nu)S_g v_s}{\pi D^3 E} = \frac{16(1 + 0.30)(2.00)(26.5)}{\pi(1)^3(30,000,000)}$$

$$= 11.70(10^{-6}) \text{ V/(in.-lb)} = 11.70 \text{ }\mu\text{V/(in-lb)}$$

Exercise 8.14

$$p_g = v_g i_g = (v_s/2)(v_s/2R_g)$$

$$v_s = 2\sqrt{p_g R_g} = 2\sqrt{0.5(1000)} = 44.7 \text{ V}$$

From Eq. (8.27):

$$S = \frac{16(1 + \nu)S_g v_s}{\pi D^3 E} = \frac{16(1 + 0.30)(2.00)(44.7)}{\pi(1)^3(30,000,000)}$$

$$= 19.73(10^{-6}) \text{ V/(in.-lb)} = 19.73 \text{ }\mu\text{V/(in-lb)}$$

Exercise 8.15

From Eq. (8.7):

$$T = \frac{(v_o/v_s)}{(v_o/v_s)^*} T_{max} = \frac{(18/8)}{(4)}(500) = 281 \text{ ft-lb}$$

Exercise 8.16

From Eq. (8.27):
$$S = \frac{16(1+\nu)S_g v_s}{\pi D^3 E} = Kv_s = \frac{v_o}{T}$$

$$\left(\frac{v_o}{v_s}\right)^* = KT_{max}$$

Therefore,
$$K = \frac{(v_o/v_s)^*}{T_{max}} = \frac{4(10^{-3})}{500} = 8.0(10^{-6}) \text{ (V/V)/(ft-lb)}$$

$$S = Kv_s = 8.0(10^{-6})(10) = 80(10^{-6}) \text{ V/(ft-lb)} = 80 \text{ } \mu\text{V/(ft-lb)}$$

Exercise 8.17

Slip rings exhibit relatively large resistance fluctuations which make it impossible to connect a gage through the rings to a quarter bridge. Instead, a full bridge is assembled on the rotating element (in this case the torque cell). Four connections are necessary for a complete bridge - two for the power supply and two for the output. With the resistance of the rings in series with the power supply and/or the readout device, small changes in ring resistance have little effect on the readout.

Exercise 8.18

The most significant advantage is when a shaft end is not free and slip rings cannot be employed. In this instance telemetry is mandatory. A second advantage is when the instruments are far from the shaft. Errors due to long lead wires can be avoided although higher power transmitters are necessary for long distance telemetry.

1. Slip ring problems are eliminated.

2. Signals from several transducers can be transmitted simultaneously.

3. Signals can be conveniently transmitted over longer distances.

4. Access to an end of the shaft is not required.

Exercise 8.19

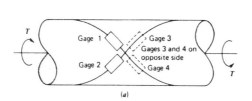
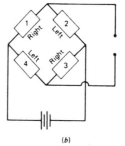

Mount four gages on the shaft and connect them into a Wheatstone bridge as shown above.

$$T = \frac{33000 \text{ hp}}{2\pi N} = \frac{33000(200)}{2\pi(800)} = 1313 \text{ ft-lb}$$

$$\tau = \frac{TR}{J} = \frac{1313(12)(1.5)}{(\pi/2)(1.5)^4} = 2972 \text{ psi (elastic)}$$

With $S_g = 2.00$, $R_g = 120\ \Omega$, and $v_s = 10$ V:

From Eq. (8.27):

$$S = \frac{16(1+\nu)S_g v_s}{\pi D^3 E} = \frac{16(1+0.30)(2.00)(10)}{\pi(3)^3(30,000,000)}$$

$$= 0.163(10^{-6}) \text{ V/(in.-lb)}$$

$$= 0.163\ \mu\text{V/(in-lb)}$$

Exercise 8.20

From Eqs. (7.48):

$$\sigma_r = \frac{E}{(1-\nu^2)}(\varepsilon_r + \nu\varepsilon_\theta) \qquad \sigma_\theta = \frac{E}{(1-\nu^2)}(\varepsilon_\theta + \nu\varepsilon_r)$$

Substituting Eqs. (8.34) yields:

$$\sigma_r = \frac{3p}{8t^2}\left[\left(R_o^2 - 3r^2\right) + \nu\left(R_o^2 - r^2\right)\right]$$

$$\sigma_\theta = \frac{3p}{8t^2}\left[\left(R_o^2 - r^2\right) + \nu\left(R_o^2 - 3r^2\right)\right]$$

At $r = 0$:
$$\sigma_r = \frac{3pR_o^2(1+\nu)}{8t^2} \qquad \sigma_\theta = \frac{3pR_o^2(1+\nu)}{8t^2} = \sigma_r$$

At $r = R_o$:
$$\sigma_r = -\frac{3pR_o^2}{4t^2} = \sigma_{max} \qquad \sigma_\theta = -\frac{3\nu pR_o^2}{4t^2} = \nu\sigma_r$$

For $p_{max} = 2000$ psi and $\sigma_{max} = S_f = 75{,}000$ psi:

$$t = \left[\frac{3pR_o}{4\sigma_{max}}\right]^{1/2} = \left[\frac{3(2000)(0.75)^2}{4(75{,}000)}\right]^{1/2} = 0.1061 \text{ in.}$$

The center point deflection of the diaphragm is given by the expression:

$$\delta_c = \frac{3pR_o^4(1-\nu^2)}{16t^3 E} = \frac{3(2000)(0.75)^4(1-0.30^2)}{16(0.1061)^3(30{,}000{,}000)} = 0.00301 \text{ in.}$$

Since $\delta_c < t/4$ the output will be linear within 0.3 percent. Use a 050 HR LVDT with a nominal linear range of ± 0.050 in., a sensitivity S_t of 6.3 (V/V)/in., and an input voltage v_s of 10 V. The LVDT is linear within 0.10 percent for 50 percent of full range.

$$S_\delta = \frac{\delta_c}{p} = \frac{0.00301}{2000} = 1.505(10^{-6}) \text{ in./psi} = 1.505 \; \mu\text{in./psi}$$

$$S = \frac{v_o}{p} = S_\delta S_t v_s = 1.505(10^{-6})(6.3)(10) = 94.8(10^{-6}) \text{ V/psi} = 94.8 \; \mu\text{V/psi}$$

Exercise 8.21

From Exercise 8.20: $\quad t = 0.1061$ in.

From Eq. (8.36):

$$S_p = 1.64 \frac{R_o^2(1 - \nu^2)\sqrt{p_T R_T}}{Et^2} = \frac{1.64(0.75)^2(1 - 0.30^2)\sqrt{1(350)}}{30{,}000{,}000(0.1061)^2}$$

$$= 46.6(10^{-6}) \text{ V/psi} = 46.6 \text{ }\mu\text{V/psi}$$

Exercise 8.22

From Exercise 8.20: $\quad t = 0.1061$ in.

From Eq. (8.38):

$$f_r = 0.471 \frac{t}{R_o^2}\left[\frac{Eg}{w(1 - \nu^2)}\right]^{1/2} = 0.471 \frac{0.1061}{(0.75)^2}\left[\frac{30(10^6)(386)}{0.285(1 - 0.30^2)}\right]^{1/2}$$

$$= 18.77(10^3) \text{ Hz} = 18.77 \text{ kHz}$$

Exercise 8.23

For a thick-walled cylinder under internal pressure:

$$\sigma_r = \frac{R_i^2 p}{R_o^2 - R_i^2}\left(1 - \frac{R_o^2}{r^2}\right) \qquad \sigma_\theta = \frac{R_i^2 p}{R_o^2 - R_i^2}\left(1 + \frac{R_o^2}{r^2}\right) \qquad \sigma_a = \frac{R_i^2 p}{R_o^2 - R_i^2}$$

For $\sigma_{max} = 80{,}000$ psi and $p = 60{,}000$ psi:

At $r = R_i$:
$$\sigma_{max} = \sigma_\theta = \frac{R_i^2 p}{R_o^2 - R_i^2}\left(1 + \frac{R_o^2}{R_i^2}\right) = \frac{R_o^2 + R_i^2}{R_o^2 - R_i^2} p = \frac{(R_o/R_i)^2 + 1}{(R_o/R_i)^2 - 1} p$$

Thus:
$$\frac{(R_o/R_i)^2 + 1}{(R_o/R_i)^2 - 1} = \frac{\sigma_{max}}{p} = \frac{80000}{60000} = 1.3333$$

which yields $\quad (R_o/R_i)^2 = 7.00 \quad$ or $\quad R_o/R_i = D_o/D_i = 2.646$

Therefore, use: $\quad D_o = 2.646$ in. $\quad$ and $\quad D_i = 1.000$ in.

Exercise 8.24

For a thick-walled cylinder under internal pressure:

$$\sigma_r = \frac{R_i^2 \, p}{R_o^2 - R_i^2}\left(1 - \frac{R_o^2}{r^2}\right) \qquad \sigma_\theta = \frac{R_i^2 \, p}{R_o^2 - R_i^2}\left(1 + \frac{R_o^2}{r^2}\right) \qquad \sigma_a = \frac{R_i^2 \, p}{R_o^2 - R_i^2}$$

On the outside surface: $r = R_o = 2.646 \, R_i$ (from Exercise 8.23)

Therefore:
$$\sigma_\theta = \frac{2R_i^2 p}{R_o^2 - R_i^2} = 0.3332 \, p \qquad \sigma_a = \frac{R_i^2 p}{R_o^2 - R_i^2} = 0.1666 \, p$$

$$\varepsilon_\theta = \frac{1}{E}(\sigma_\theta - \nu\sigma_a)$$

$$= \frac{1}{30(10^6)}\left[0.3332 \, p - 0.30(0.1666 \, p)\right] = 9.44(10^{-9}) \, p$$

From Eq. (5.5):

$$\frac{\Delta R}{R} = S_g \varepsilon_\theta = 2.00(9.44)(10^{-9})p = 18.88(10^{-9})p$$

$$p_g = v_g i_g = (v_s/2)(v_s/2R_g)$$

$$v_s = 2\sqrt{p_g R_g} = 2\sqrt{0.5(350)} = 26.5 \text{ V}$$

With active gages in positions 1 and 3 and dummy gages in positions 2 and 4 of a Wheatstone bridge:

$$v_o = \frac{1}{4}\left[\frac{\Delta R_1}{R_1} + \frac{\Delta R_3}{R_3}\right] v_s$$

$$= \frac{1}{4}(2S_g \varepsilon_\theta)v_s = \frac{1}{4}(2)[18.88(10^{-9})p](26.5) = 0.250(10^{-6})p$$

$$S = \frac{v_o}{p} = 0.250(10^{-6}) \text{ V/psi} = 0.250 \, \mu\text{V/psi}$$

Exercise 8.25

At the positions shown, the gages experience no strain but can provide temperature compensation. They also serve as bridge completion resistors.

Exercise 8.26

For a twisting moment $M_z = T$, $\varepsilon_x = \varepsilon_y = 0$; therefore, the combined axial force and moment transducer which has gages oriented only in the x-direction is not sensitive to a torque M_z.

Exercise 8.27

From Eq. (8.48):

$$\varepsilon_{max} = \frac{2.0}{\omega_n t_o} = \frac{2.0}{2\pi f_n t_o} = \frac{2.0}{2\pi(10)(10^3)(0.001)} = 0.0318 = 3.18\%$$

Exercise 8.28

From Eq. (8.48):

With $f = 20$ kHz:

$$\varepsilon_{max} = \frac{2.0}{\omega_n t_o} = \frac{2.0}{2\pi f_n t_o} = \frac{2.0}{2\pi(20)(10^3) t_o} = \frac{15.92(10^{-6})}{t_o}$$

t_o (s)	ε_{max} (%)
0.0001	15.9155
0.001	1.5915
0.01	0.1592
0.1	0.0159
1.0	0.0016

Exercise 8.29

From Eq. (8.55):
$$\mathcal{E} = \frac{(\omega/\omega_n)^2}{1 - (\omega/\omega_n)^2}$$

(a) For $\omega/\omega_n = 0.05$:
$$\mathcal{E} = \frac{(0.05)^2}{1 - (0.05)^2} = 0.0025 = 0.25\%$$

Case	ω/ω_n	$\mathcal{E}_{max}$ (%)
(a)	0.05	0.25
(b)	0.10	1.01
(c)	0.20	4.17
(d)	0.50	33.33

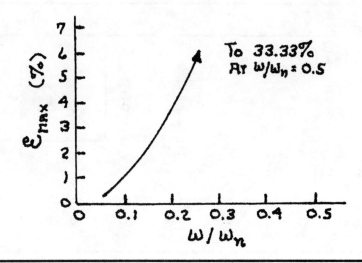

To 33.33% At $\omega/\omega_n = 0.5$

Exercise 8.30

From Eq. (8.40):
$$m\ddot{x} + C\dot{x} + kx = \dot{F}_0 t$$

where $\dot{F}_0 = F_0/t_0$, $C = 2dk/(\sqrt{k/m})$, and $k = P/\delta$. Since $C = 0$ when $d = 0$,

$$m\ddot{x} + kx = \frac{F_0 t}{t_0}$$

Which has the solution:

$$x = \frac{F_0}{k(\sqrt{k/m})t} \sin(\sqrt{k/m})t + \frac{F_0 t}{kt_0} = \frac{\dot{F}_0}{k\omega_n} \sin \omega_n t + \frac{\dot{F}_0 t}{k}$$

Therefore:
$$\frac{xk}{\dot{F}_0} = \frac{1}{\omega_n} \sin \omega_n t + t$$

Exercise 8.31

From Exercise 8.30:
$$\frac{xk}{\dot{F}_0} = \frac{1}{\omega_n} \sin \omega_n t + t$$

which is a sinusoidal oscillation about the ramp function $F(t) = \dot{F}_0 t$.

Since $F_{measured} = kx$:
$$F_{measured} = \frac{\dot{F}_o}{\omega_n} \sin \omega_n t + \dot{F}_o t$$

$$\mathcal{E} = \frac{F_{measured} - \dot{F}_o t}{\dot{F}_o t} = \frac{\sin \omega_n t}{\omega_n t}$$

$\omega_n t$	$\mathcal{E}$ (%)	$\omega_n t$	$\mathcal{E}$ (%)	$\omega_n t$	$\mathcal{E}$ (%)
0	100	$5\pi/2$	12.7	5π	0
$\pi/2$	64	3π	0	$11\pi/2$	-5.8
π	0	$7\pi/2$	-9.1	6π	0
$3\pi/2$	-21	4π	0	$13\pi/2$	4.9
2π	0	$9\pi/2$	7.1	7π	0

The sinusoidal oscillation decreases in importance as the magnitude of the output increases.

Exercise 8.32

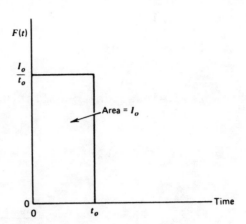

From Eq. (8.40):

$$m\ddot{x} + C\dot{x} + kx = F_0$$

where $C = 2dk/(\sqrt{k/m})$, and $k = P/\delta$.

For $0 \le t \le t_0$: $\quad x_c = e^{-at}(A\cos bt + B\sin bt)$

where: $\quad a = d\omega_n \quad$ and $\quad b = \omega_n\sqrt{1-d^2}$

For $0 \le t \le t_0$: $\quad x_p = \dfrac{F_0}{k} = \dfrac{I_0}{kt_0}$

Therefore, $\quad x = x_c + x_p$

$$= e^{-at}(A\cos bt + B\sin bt) + \dfrac{F_0}{k}$$

Initial conditions $\quad x = 0$ at $t = 0 \quad$ yields $\quad A = -F_0/k$

$\dot{x} = 0$ at $t = 0 \quad$ yields $\quad B = \dfrac{a}{b}A$

$$x = \dfrac{F_0}{k}\left[1 - e^{-at}\left(\cos bt + \dfrac{d}{\sqrt{1-d^2}}\sin bt\right)\right]$$

$$= \dfrac{F_0}{k}\left[1 - \dfrac{e^{-d\omega_n t}}{\sqrt{1-d^2}}\cos\left(\omega_n\sqrt{1-d^2} - \phi\right)\right]$$

where $\quad \tan\phi = \dfrac{d}{\sqrt{1-d^2}}$

The above equations are valid for $0 < t < t_0$. For $t > t_0$ superimpose a negative impulse that is time shifted by an amount t_0 to obtain the solution $x(t - t_0)$.

Exercise 8.33

From Exercise 8.32:

$$x = \frac{F_o}{k}\left[1 - \frac{e^{-d\omega_n t}}{\sqrt{1-d^2}} \cos\left(\omega_n \sqrt{1-d^2} - \phi\right)\right]$$

where

$$\tan \phi = \frac{d}{\sqrt{1-d^2}}$$

d	$\sqrt{1-d^2}$	$1/\sqrt{1-d^2}$	$\phi°$	ϕ rad
0	1	1	0	1
0.01	0.9999	1.0001	0.57	0.0032π
0.10	0.9950	1.0050	5.74	0.0319π
0.20	0.9798	1.0206	11.5	0.0639π
0.50	0.8660	1.1547	30.0	0.1667π
0.70	0.7141	1.4003	44.4	0.2467π

A typical response curve for the case d = 0.05 follows.

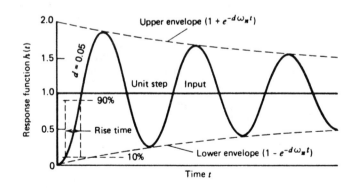

ENGINEERING MEASUREMENTS by J. W. DALLY, W. F. RILEY, AND K. G. McCONNELL

Exercise 8.34

Transducers with accuracies of 0.1 percent are available commercially. Because the calibration constant is stable over long periods of time, recalibration every 6 to 12 months is sufficient. Large changes in the initial zero or a significant change in the calibration constant usually implies that the transducer has been overloaded or abused in some other way.

Errors result from improper calibration, from abuse, and from excessively long periods between recalibration. Errors also occur when transducers are used at only a small fraction of their range. If a 10,000 lb load cell is used to measure 100 lbs, the 0.1 % accuracy increases by a factor of (10,000/100) to 10 %.

Errors also occur when transducers are used in dynamic experiments where the excitation frequencies approach or exceed the natural frequency of the mechanical element in the transducer.

Exercise 8.35

The specification should include the following paragraphs:

(a) For pressure transducers:

1. Discuss the importance of calibration and period for recalibration.
2. Discuss dead-weight calibration.
3. Specify incremental weights and calibration pressures.
4. Describe a method for preparing a graph of pressure p versus v_o.
5. Discuss use of a least squares fit to determine the calibration constant.

(b) For load cells:

1. Discuss the importance of calibration and period for recalibration.
2. Discuss dead-weight calibration for low-capacity load cells.
3. Discuss calibration with a testing machine of certified accuracy.
4. Discuss calibration with a "standard" load cell.
5. Describe a method for preparing a graph of load P versus v_o.
6. Discuss use of a least squares fit to determine the calibration constant.

(c) For torque cells:

1. Discuss the importance of calibration and period for recalibration.
2. Discuss dead-weight calibration for low-capacity torque cells.
3. Discuss calibration with a testing machine of certified accuracy.
4. Discuss calibration with a "standard" torque cell.
5. Describe a method for preparing a graph of torque T versus v_o.
6. Discuss use of a least squares fit to determine the calibration constant.

Exercise 9.1

Recall that
$$\cos(x \pm y) = \cos x \cos y \mp \sin x \sin y$$
$$\sin(x \pm y) = \cos x \sin y \pm \sin y \cos x$$

For:
$$x = 8\cos(12t) + 6\sin(12t)$$
$$= A\cos(12t + \theta_1)$$
$$= A[\cos(12t)\cos\theta_1 - \sin(12t)\sin\theta_1]$$

Therefore:
$$\left.\begin{array}{l} A\sin\theta_1 = -6 \\ A\cos\theta_1 = +8 \end{array}\right\} \quad A = \sqrt{6^2 + 8^2} = 10$$
$$\theta_1 = \tan^{-1}\frac{-6}{+8} = -36.8° = -0.644 \text{ rad}$$

For:
$$x = 8\cos(12t) + 6\sin(12t)$$
$$= B\sin(12t + \theta_2)$$
$$= B[\cos(12t)\sin\theta_2 + \sin(12t)\cos\theta_2]$$

Therefore:
$$\left.\begin{array}{l} B\sin\theta_2 = +8 \\ B\cos\theta_2 = +6 \end{array}\right\} \quad B = \sqrt{6^2 + 8^2} = 10$$
$$\theta_2 = \tan^{-1}\frac{8}{6} = +53.1° = +0.927 \text{ rad}$$

θ_1 is lagging (- sign): $x = A\cos(12t + \theta_1) = 10\cos(12t - 0.644)$

θ_2 is leading (+ sign): $x = B\sin(12t + \theta_2) = 10\sin(12t + 0.927)$

Reference is cosine for phase angle θ_1 and sine for phase angle θ_2.

Exercise 9.2

$$\omega = 2\pi f = 2\pi(100) = 628 \text{ rad/s}$$
$$V = \delta\omega = 0.001(628) = 0.628 \text{ in./s} = V_{max}$$
$$a = \delta\omega^2/g = 0.001(628)^2/386 = 1.02 \text{ g's} = a_{max}$$

With $\delta_1 = 2\delta$ and $f_1 = f$ Both V and a are doubled.

With $\delta_2 = \delta$ and $f_2 = 2f$ V is doubled and a is quadrupled.

Exercise 9.3

From Eq. (9.6):
$$H(\omega) = \frac{1}{k - m\omega^2 + jC\omega} = \frac{1}{k[(1 - r^2) + j2rd]}$$

where $r = \omega/\omega_n$, $\omega_n = \sqrt{k/m}$, and $d = C/(2\sqrt{km})$.

$$H(\omega) = \frac{(1 - r^2) - j(2rd)}{k[(1 - r^2)^2 + (2rd)^2]} = \text{Re}[H(\omega)] + \text{Im}[H(\omega)]$$

For $k = 1.0(10^6)$ lb and $d = 0.02$:

r	$1-r^2$	2rd	$[(1-r^2)^2 + (2rd)^2]$	Re[H(ω)]	Im[H(ω)]
0	1	0	1	$1.00(10^{-6})$	0
0.75	0.4375	0.0300	0.1923	$2.28(10^{-6})$	$-0.16(10^{-6})$
0.97	0.0591	0.0388	0.004999	$11.8(10^{-6})$	$-7.76(10^{-6})$
1.00	0	0.0400	0.0016	0	$-25.0(10^{-6})$
1.03	-0.0609	0.0412	0.005406	$-11.3(10^{-6})$	$-7.63(10^{-6})$
1.25	-0.5625	0.0500	0.3189	$-1.76(10^{-6})$	$-0.16(10^{-6})$
2	-3	0.0800	9.0064	$-0.33(10^{-6})$	$-0.01(10^{-6})$

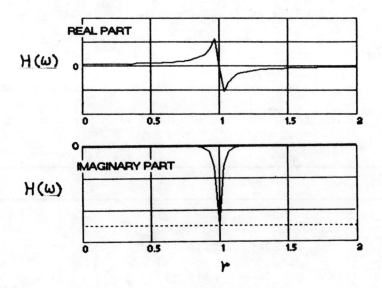

(a) Observe in the above plots that the real part starts at a value $1/k$ at $r = 0$, peaks before $r = 1.0$, passes through zero at $r = 1.0$, has a negative trough after $r = 1.0$, and approaches zero as $r \rightarrow 2$.

(b) The imaginary part starts at zero, reaches a maximum negative value of $1/(2dk)$ at $r = 1.0$, and returns to zero as $r \rightarrow 2$.

(c) Figure 9.3 shows the magnitude and phase in a normalized plot.

Exercise 9.4

The equation of motion for the uniform bar of mass m_b and length ℓ shown in Fig. E9.4 is obtained by summing moments about point O. Thus,

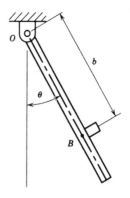

$$+\circlearrowleft \Sigma M_O = -m_b g \frac{\ell}{2} \sin\theta = I_O \ddot{\theta}$$

or

$$I_O \ddot{\theta} + m_b g \frac{\ell}{2} \sin\theta = 0$$

For a slender bar of uniform cross section,

$$I_O = \frac{1}{3} m_b \ell^2$$

Therefore,

$$\ddot{\theta} = -\frac{3g}{2\ell} \sin\theta$$

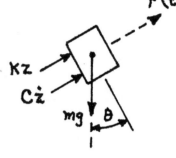

At point B:

$$A_B = b\ddot{\theta} = -\frac{3b}{2\ell} g \sin\theta$$

The equation of motion for the seismic mass is given by Eq. (9.3) as

$$m\ddot{z} + C\dot{z} + kz = F(t) - m\ddot{x}$$

From the free-body diagram of the seismic mass:

$$F(t) = -mg \sin\theta$$

Therefore,

$$m\ddot{z} + C\dot{z} + kz = -mg \sin\theta - mA_B = -mg \sin\theta \left[1 - \frac{3b}{2\ell}\right]$$

Note that the exciting force $F(t) = 0$ when $b = 2\ell/3$ and $F(t) = -mg \sin\theta$ when $b = 0$. This occurs because the gravitational force of attraction on the seismic mass changes as the orientation of the accelerometer's sensitive axis varies with angular position θ.

ENGINEERING MEASUREMENTS by J. W. DALLY, W. F. RILEY, AND K. G. McCONNELL

Exercise 9.5

For the seismic instrument:

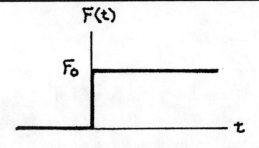

$$m\ddot{z} + C\dot{z} + kz = F(t) = F_0$$

which has the solution

$$z = z_p + z_t$$

where
$$z_p = F_0/k \quad \text{(particular solution)}$$

$$z_t = Be^{-d\omega_n t} \cos(\omega_d t - \phi) \quad \text{(transient solution)}$$

$$z = \frac{F_0}{k} + Be^{-d\omega_n t} \cos(\omega_d t - \phi)$$

From the initial condition:

$z = 0$ at $t = 0$: $\quad \dfrac{F_0}{k} + B\cos(-\phi) = 0$

$\dot{z} = 0$ at $t = 0$: $\quad B[-d\omega_n \cos(-\phi) - \omega_d \sin(-\phi)] = 0$

Thus: $\quad \tan\phi = d\omega_n/\omega_d = d/(1 - d^2)^{1/2}$

$\cos\phi = (1 - d^2)^{1/2}$

Therefore: $\quad B = -\dfrac{F_0}{k(1 - d^2)^{1/2}}$

and $\quad z(t) = \dfrac{F_0}{k}\left[1 - \dfrac{e^{-d\omega_n t}}{(1 - d^2)^{1/2}} \cos(\omega_d t - \phi)\right]$

This response, as shown in the figure below, decays exponentially to the final value of (F_0/k) while Fig. 9.4 shows no decay.

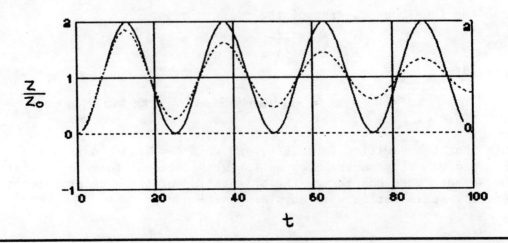

Exercise 9.6

$$\omega_n = 2\pi f_n = \sqrt{k/m}$$

$$k = (2\pi f_n)^2 m = 4\pi^2(5)^2(0.20/386) = 0.511 \text{ lb/in.}$$

From Eqs. (9.8), (9.9), and (9.15):

$$v_o = \frac{r^2}{\sqrt{(1-r^2)^2 + (2rd)^2}} S_v V_0 \quad \text{and} \quad \phi = \tan^{-1} \frac{2rd}{1-r^2}$$

For $V_0 = 8.0$ in./s and $f = 8.0$ Hz:

$$r = f/f_n = 8.0/5.0 = 1.6$$
$$2rd = 2(1.6)(0.05) = 0.16$$
$$1 - r^2 = 1 - (1.6)^2 = -1.56$$

$$v_o = \frac{(1.6)^2}{\sqrt{(-1.56)^2 + (0.16)^2}}(8.3)(8.0) = 108.4 \text{ mV}$$

$$\phi = \tan^{-1} \frac{2rd}{1-r^2} = \tan^{-1} \frac{0.16}{-1.56} = 174.1° = 3.04 \text{ rad}$$

For $V_0 = 8.0$ in./s and $f = 20.0$ Hz:

$$r = f/f_n = 20.0/5.0 = 4.0$$
$$2rd = 2(4.0)(0.05) = 0.40$$
$$1 - r^2 = 1 - (4.0)^2 = -15.0$$

$$v_o = \frac{(4.0)^2}{\sqrt{(-15.0)^2 + (0.40)^2}}(8.3)(8.0) = 70.8 \text{ mV}$$

$$\phi = \tan^{-1} = \frac{2rd}{1-r^2} = \tan^{-1} \frac{0.40}{-15.0} = 178.5° = 3.11 \text{ rad}$$

Ideally: $\quad v_o = S_v V_0 = 8.3(8.0) = 66.4$ mV

At 8 Hz: $\quad \mathcal{E} = \dfrac{108.4 - 66.4}{66.4} = 0.633 = 63.3\%$ high (Near resonance)

At 20 Hz: $\quad \mathcal{E} = \dfrac{70.8 - 66.4}{66.4} = 0.0663 = 6.63\%$ high

The error at 8 Hz could be reduced by increasing the damping d. A second coil and resistor could be used to induce additional damping.

Exercise 9.7

$$a(t) = a \sin(0.2\omega_n t) - 0.4a \sin(0.6\omega_n t)$$

$$z(t) = b_1 \sin(0.2\omega_n t - \phi_1) + b_3 \sin(0.6\omega_n t - \phi_3)$$

$$b_i = \frac{a_i}{\sqrt{(1-r_i^2)^2 + (2r_i d)^2}} \quad \text{and} \quad \phi_i = \tan^{-1} = \frac{2r_i d}{1 - r_i^2}$$

For $d = 0.05$:

Freq Comp	r	$1 - r^2$	2rd	H	b	tan ϕ	ϕ	ϕ_{lin}
1	0.2	0.96	0.020	1.041	1.041a	0.0208	1.19	18.0
3	0.6	0.64	0.060	1.556	-0.622a	0.0938	5.36	54.0

For $d = 0.60$:

Freq Comp	r	$1 - r^2$	2rd	H	b	tan ϕ	ϕ	ϕ_{lin}
1	0.2	0.96	0.240	1.011	1.011a	0.250	14.0	18.0
3	0.6	0.64	0.720	1.038	-0.415a	1.125	48.4	54.0

The 60 % damping gives the best results with amplitude distortion under 4 % and phase distortions of 22 % and 10 %.

Exercise 9.8

$$v_1 = \frac{q_{A/C}}{C_1} = \frac{k_1 a_y - k_2 \alpha_z}{C_1} = S_1 a_y - S_2 \alpha_z$$

$$v_2 = \frac{q_{B/D}}{C_2} = \frac{k_3 a_y + k_4 \alpha_z}{C_2} = S_3 a_y + S_4 \alpha_z$$

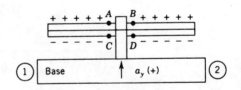

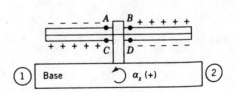

$$v_a = v_1 + v_2 = (S_1 + S_3)a_y + (S_4 - S_2)\alpha_z \cong S_a a_y$$

$$v_\alpha = v_2 - v_1 = (S_3 - S_1)a_y + (S_4 + S_2)\alpha_z \cong S_\alpha \alpha_z$$

The effects are the same as in other transducers; i.e., same mechanical limitations.

Exercise 9.9

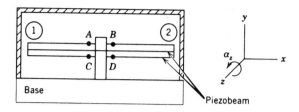

For $V_o = 2.0$ in./s and $\dot{\theta} = 6.0$ rad/s at 100 Hz:

$$\ddot{\theta} = \omega\dot{\theta} = 2\pi f\dot{\theta} = 2\pi(100)(6.0) = 3770 \text{ rad/s}^2 = \alpha$$

For $S_\alpha = 0.5$ mV/(rad/s^2): $\quad v_{max} = S_\alpha \alpha = 0.5(10^{-3})(3770) = 1.885$ V

For $S_\alpha = 5$ mV/(rad/s^2): $\quad v_{max} = S_\alpha \alpha = 5(10^{-3})(3770) = 18.85$ V

For $S_\alpha = 50$ mV/(rad/s^2): $\quad v_{max} = S_\alpha \alpha = 50(10^{-3})(3770) = 188.5$ V

Select $S_\alpha = 0.5$ mV/(rad/s^2) due to ±10 V limit for many common solid state amplifiers.

Exercise 9.10

$$C = C_t + C_c + C_a = 1000 + 312 + 15 = 1327 \text{ pF}$$

$$C_{eq} = \frac{CC_b}{C + C_b} = \frac{1327(10,000)}{1327 + 10,000} = 1172 \text{ pF}$$

(a) From Eq. (9.22):

$$S_v = \frac{Sq^*}{C} = \frac{83}{1327} = 62.5(10^{-3}) \text{ V/g} = 62.5 \text{ mV/g}$$

(b) $\quad \mathcal{E} = \dfrac{\tau - \tau_{eq}}{\tau} = \dfrac{RC - RC_{eq}}{RC} = \dfrac{C - C_{eq}}{C} = \dfrac{1327 - 1172}{1327} = 0.1168 = 11.68\%$

(c) At the break frequency: $\quad RC_{eq}\omega = RC_{eq}(2\pi f) = 1$

$$f = \frac{1}{2\pi RC_{eq}} = \frac{1}{2\pi(100)(10^6)(1172)(10^{-12})} = 1.358 \text{ Hz}$$

Exercise 9.11

From Table 9.1 for a Kistler 808A accelerometer:

$$C_t = 90 \text{ pF} \quad \text{and} \quad S_q^* = 1.0 \text{ pC/g}$$

$$C = C_t + C_c = 90 + 10(30) = 390 \text{ pF}$$

(a) For $\mathcal{E} \leq 1\%$:
$$C_{eq} = \frac{CC_b}{C + C_b} \cong 0.99C \quad \text{or} \quad \frac{C_b}{C + C_b} \cong 0.99$$

Therefore, $C_b \geq 99C \geq 99(390) \geq 38{,}610 \text{ pF} \geq 40 \text{ nF}$

(b) From Eq. (9.22):

$$S_v = \frac{S_q^*}{C} = \frac{1.0(10^{-12})}{390(10^{-12})} = 2.56(10^{-3}) \text{ V/g} = 2.56 \text{ mV/g}$$

$$G = \frac{S_{v0}}{S_v} = \frac{10.0}{2.56} = 3.91$$

(c) $\tau = RC_{eq} = 1000(10^6)(390)(10^{-12}) = 0.390 \text{ s}$

Exercise 9.12

From Eq. (9.26): $C_{eq} = C_f\left(1 + \dfrac{C}{G_1 C_f}\right)$ where $\dfrac{C}{G_1 C_f}$ is the error.

From Table 9.4: $C = 0.001 \text{ } \mu\text{F}$ for $C_f = 10 \text{ pF}$ for 0.5% error.

$C = 0.1 \text{ } \mu\text{F}$ for $C_f = 1000 \text{ pF}$ for 0.5% error.

(a) $G = \dfrac{C}{0.005 C_f} = \dfrac{0.001(10^{-6})}{0.005(10)(10^{-12})} = 20{,}000$

(b) $G = \dfrac{C}{0.005 C_f} = \dfrac{0.1(10^{-6})}{0.005(1000)(10^{-12})} = 20{,}000$

(c) If C_{cal} is grounded, the source capacitance is increased by 10,000 pF. Thus $C \cong 10{,}000 \text{ pF} \cong 0.010 \text{ } \mu\text{F}$. Assuming $G_1 = 20{,}000$ and $C_f = 10 \text{ pF}$,

$$\mathcal{E} = \frac{C}{G_1 C_f} = \frac{0.010(10^{-6})}{20{,}000(10)(10^{-12})} = 0.05 = 5\%$$

Exercise 9.13

From Eq. (9.28):
$$S_v = \frac{S_q^*}{b\, C_f}$$

Using subscripts m, s, and t for measured, set, and true:

$$v_m = S_v a = \frac{S_q^*}{b_s C_f} a \quad \text{or} \quad v_m b_s = v_t b_t = \frac{S_q^*}{C_f} a$$

Therefore:
$$v_t = \frac{b_s}{b_t} v_m = \frac{4.86}{8.46} v_m = 0.5745\, v_m$$

Exercise 9.14

$$C = C_t + C_c = 10{,}000 + 30(30) = 10{,}900 \text{ pF}$$

From Eq. (9.26):
$$C_{eq} = C_f\left(1 + \frac{C}{G_1 C_f}\right) \quad \text{where} \quad \frac{C}{G_1 C_f} \text{ is the error.}$$

For $\varepsilon \leq 0.5\%$:
$$\frac{C}{G_1 C_f} \leq 0.005$$

Therefore:
$$C_f = \frac{C}{0.005\, G_1} = \frac{10{,}900(10^{-12})}{0.005(20{,}000)} = 109(10^{-12})\text{ F} = 109 \text{ pF}$$

(a) $\quad S_a = \dfrac{1}{S_v} = \dfrac{b\, C_f}{S_q^*} = \dfrac{0.085(200)(10^{-12})}{170(10^{-12})} = 0.100 \text{ g/V}$

(b) $\quad S_a = \dfrac{1}{S_v} = \dfrac{b\, C_f}{S_q^*} = \dfrac{0.170(200)(10^{-12})}{170(10^{-12})} = 0.200 \text{ g/V}$

(c) $\quad S_a = \dfrac{1}{S_v} = \dfrac{b\, C_f}{S_q^*} = \dfrac{0.340(200)(10^{-12})}{170(10^{-12})} = 0.400 \text{ g/V}$

The $b = 0.085$ setting gives the largest voltage per unit of acceleration.

ENGINEERING MEASUREMENTS by J. W. DALLY, W. F. RILEY, AND K. G. McCONNELL

Exercise 9.15

Let $u = \tau\omega$ and $u_1 = \tau_1\omega = nu = n\tau\omega$

From Eq. (9.31): $|H(\omega)| = \dfrac{nu^2}{\sqrt{(1+u^2)(1+n^2u^2)}} = \alpha$

$$u^2 = \dfrac{\alpha^2(1+n^2) + \sqrt{\alpha^4(1+n^2)^2 + 4\alpha^2 n^2(1-\alpha^2)}}{2n^2(1-\alpha^2)}$$

For $\alpha = 0.95$ and $n = 1$: $\quad u^2 = 19.00 \quad\quad u = 4.36$

For $\alpha = 0.95$ and $n = 10$: $\quad u^2 = 9.351 \quad\quad u = 3.06$

For $\alpha = 0.95$ and $n = 100$: $\quad u^2 = 9.256 \quad\quad u = 3.04$

$\tau_1 = 10\tau$ is nearly the same as a single time constant system.

Exercise 9.16

From Table 9.6 for a rectangular pulse with $t_1 = 4$ ms and 5% error:

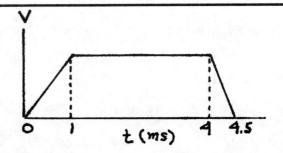

$\tau = RC = 20t_1 = 20(4.0) = 80$ ms (minimum)
$\quad\quad\quad\quad\quad = 20(4.5) = 90$ ms (preferred)

For a ramp-hold response:

$$\varepsilon_{max} = \dfrac{2}{\omega_n t_0} = \dfrac{2}{2\pi f_n t_0}$$

$$f_n = \dfrac{2}{2\pi\varepsilon_{max} t_0} = \dfrac{2}{2\pi(0.050)(1)(10^{-3})}$$

$$= 6.37(10^3) \text{ Hz} = 6.37 \text{ kHz}$$

Exercise 9.17

$$v = \frac{S_v A_0}{\tau - \tau_1}\left(\tau e^{-t/\tau_1} - \tau_1 e^{-t/\tau}\right)$$

$$\frac{dv}{dt} = \frac{S_v A_0}{\tau - \tau_1}\left(-\frac{\tau}{\tau_1} e^{-t/\tau_1} + \frac{\tau_1}{\tau} e^{-t/\tau}\right)$$

Evaluating at t = 0 gives the initial slope as:

$$\left.\frac{dv}{dt}\right]_{t=0} = -\frac{S_v A_0}{\tau - \tau_1}\left(\frac{\tau^2 - \tau_1^2}{\tau_1 \tau}\right) = -S_v A_0 \left(\frac{\tau + \tau_1}{\tau \tau_1}\right)$$

For $\quad v = S_v A_0 e^{-t/\tau_e} \qquad \left.\frac{dV}{dt}\right]_{t=0} = -\frac{S_v A_0}{\tau_e}$

Therefore: $\qquad \tau_e = \dfrac{\tau \tau_1}{\tau + \tau_1}$

For $\tau_1 = \tau$: $\qquad \tau_e = \dfrac{1}{2}\tau$

For $\tau_1 = 10\tau$: $\qquad \tau_e = \dfrac{10}{11}\tau$

Exercise 9.18

From Table 9.1 for a PCB 302A accelerometer:

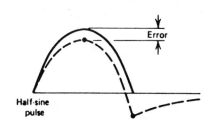

Half-sine pulse

$\tau = 0.5$ s

$S_v = 10$ mv/g

$\tau_1 = R_1 C_1 = 1(10^6)(1)(10^{-6}) = 1.0$ s

$\tau_e = \dfrac{\tau_1 \tau}{\tau + \tau_1} = \dfrac{1.0(0.5)}{1.5} = 0.333$ s $\qquad \dfrac{\tau_e}{t_1} = \dfrac{0.333}{0.1} = 3.33$ s

From Table 9.6: $\qquad$ Since $\tau_e/t_1 > 3t_1$ peak error $\mathcal{E} \cong 10\%$

$\qquad\qquad\qquad\qquad$ Since $\tau_e/t_1 > 3t_1$ undershoot error $\mathcal{E} \cong 20\%$

Exercise 9.19

From Fig. E9.19:

$2\delta = 0.1781 - 0.1419 - 0.0011$

$\quad = 0.0351$ in.

$\delta = 0.01755$ in.

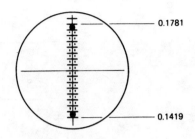

From Eq. 9.37:

$$a_{max} = \omega^2 \delta = (2\pi f)^2 \delta$$

$$\frac{a}{g} = \frac{4\pi^2(0.01755)}{386} f^2 = 1.795(10^{-3}) f^2$$

case	freq	a/g
(a)	50 Hz	4.49
(b)	100 Hz	17.9
(c)	500 Hz	449
(d)	1000 Hz	1795

$$F = ma = \frac{W}{g} a$$

With $W = 0.35$ lb and $F_{max} = 100$ lb:

$$a_{max} = \frac{F_{max}}{W} g = \frac{100}{0.35} g = 286 \, g$$

$$\delta = 0.01755 \pm 0.0001$$

$$\mathcal{E} = \frac{0.0001}{0.01755} = 0.0057 = 0.57\%$$

Exercise 9.20

Summing moments about the support yields:

$+\circlearrowleft \Sigma M_0 = I_0 \ddot{\theta}$

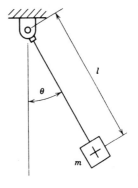

For small oscillations:

$-mg\ell\,\theta = m\ell^2\,\ddot{\theta}$

$$\ddot{\theta} + (g/\ell)\,\theta = \ddot{\theta} + \omega_n^2\,\theta = 0$$

Thus:
$$g = \omega_n^2 \ell = 4\pi^2(\ell/T^2)$$

$$dg = 4\pi^2\left(\frac{d\ell}{T^2} - \frac{2\ell\,dT}{T^3}\right)$$

$\dfrac{dg}{g} = \dfrac{d\ell}{\ell} - \dfrac{2dT}{T}$ (dT/T must be measured twice as accurately as $d\ell/\ell$).

For $\ell = 25$ in.:
$$T_o = 2\pi\sqrt{\ell/g} = 2\pi\sqrt{\frac{25}{386}} = 1.599 \approx 1.60 \text{ s.}$$

Uncertainty:
$$\frac{dT}{T} = \frac{0.15}{1.60} = 0.094 = 9.4\ \%$$

1. Use a spherical pendulum body to insure accurate measurement of ℓ.
2. Use 10 periods to reduce uncertainty to 0.94 %.
3. Doubling the pendulum length reduces $d\ell/\ell$ uncertainty by a factor of two.
4. Plotting T^2 vs ℓ gives a slope of $(4\pi^2/g)$. This helps to reduce ℓ uncertainty.

Exercise 9.21

From Eq. 9.40:
$$S_a = \left(\frac{v_a}{v_f}\right) v_{mg}$$

Test 1. $\quad S_a = \left(\dfrac{4.23}{19.48}\right) 46.6 = 10.12 \text{ mV/g}$

Test 2. $\quad S_a = \left(\dfrac{16.83}{78.1}\right)(46.5) = 10.02 \text{ mV/g}$

Test 3. $\quad S_a = \left(\dfrac{64.5}{302}\right)(46.7) = 9.97 \text{ mV/g}$

Test 4. $\quad S_a = \left(\dfrac{254}{1162}\right)(46.5) = 10.16 \text{ mV/g}$

$$S_a = \Sigma S_a / 4 = 10.07 \text{ mV/g}$$

$$V_{mg} = S_f \, mg = S_f \, \eta \, mg_0$$

$$\eta = \frac{g}{g_0} = \frac{31.30}{32.17} = 0.973$$

$$S_a = \left(\frac{v_a}{v_f}\right) v_{mg} = \left[\frac{v_a}{v_f} v_{mg_0}\right]\eta = S_a(\text{true})\,\eta$$

$$S_a(\text{true}) = \frac{S_a}{\eta} = \frac{10.07}{0.973} = 10.35 \text{ mV/g}$$

Exercise 9.22

$s = 8.06$ g/lb

$S_a = 6.20$ mV/g

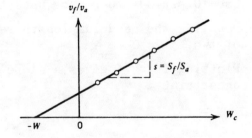

(a) From Eq. (9.44):

$$S_f = s S_a = 8.06(6.20) = 49.97 \text{ mV/lb} \cong 50.0 \text{ mV/lb}$$

(b) From the graph above:

$$s W_s = 0.403$$

$$W_s = \frac{0.403}{s} = \frac{0.403}{8.06} = 0.050 \text{ lb}$$

Exercise 9.23

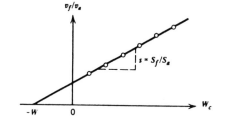

$b_a = 0.276$ (accelerometer)

$b_f = 1.00$ (force gage)

$C_{fa} = 100$ pF $C_{fF} = 2000$ pF

$S_{qa} = 2.76$ pC/g $s = 7.82$ g/lb

From Eq. 9.44:

$$S_F = sS_a = s\frac{S_{qa}}{b_a C_{fa}} = 7.82\,\frac{2.76(10^{-12})}{0.276(100)(10^{-12})} = 0.782 \text{ V/lb}$$

$$S_{qF} = S_F b_f C_{fF} = 0.782(1.00)(2000)(10^{-12})$$

$$= 1564(10^{-12}) \text{ C/lb} = 1564 \text{ pC/lb}$$

Exercise 9.24

When the accelerometer measures $\ddot{y}$ instead of $\ddot{x}$, the differential equation for the force transducer becomes:

$$C\dot{z} + kz = m\ddot{y}$$

where m includes the calibration mass m_c, the accelerometer mass m_a, and the seismic mass m_s. The output signals v_f and v_a from the force transducer and the accelerometer are:

$$v_f = S_f F = S_F \frac{W + W_c}{g} \ddot{y} \qquad \text{and} \qquad v_a = S_a \frac{\ddot{y}}{g}$$

Therefore: $\qquad v_f = \frac{S_f}{S_a}(W + W_c)v_a \qquad$ or $\qquad \frac{v_f}{v_a} = s(W + W_c)$

The calibration equation is the same. In this case, force transducer resonance does not enter into the measurement like it does in the set up shown in Fig. 9.22. However, accelerometer cable dynamics can influence the results in this case.

ENGINEERING MEASUREMENTS by J. W. DALLY, W. F. RILEY, AND K. G. McCONNELL

Exercise 9.25

$$S_a = 7.30 \text{ mV/g} \qquad W_h = 0.165 \text{ lb}$$
$$s = 7.19 \text{ g/lb} \qquad W_b = 0.769 \text{ lb}$$

From Eq. (9.50): $\quad \dfrac{v_f}{v_a} = \dfrac{S_f^*}{S_a} W_p$

Therefore: $\quad S_f^* = s S_a = 7.19(7.30) = 52.5 \text{ mV/lb}$

With $W_h = 0.611$ lb and $W_b = 1.278$ lb:

$$M = \frac{m_h}{m_b} = \frac{0.611/32.2}{1.278/32.2} = 0.478$$

Therefore: $\quad S_f^* = \dfrac{S_f}{1 + M} = \dfrac{63.8}{1.478} = 43.2 \text{ mV/lb}$

Exercise 9.26

$$S_q = 1.46 \text{ pC/psi} \qquad C_{cal} = 1000 \text{ pF} \qquad S_v = 5 \text{ mV/psi}$$

From Eq. 9.28: $\quad S_v = \dfrac{S_{qf}}{bC_f} = \dfrac{1.46(10^{-12})}{bC_f} = 0.005$

(a) Set b at: $\quad b = 0.1460$

(b) Then: $\quad C_f = \dfrac{1.46(10^{-12})}{0.1460(0.005)} = 2000 \text{F} = 2000 \text{ pF}$

(c) $\quad q_{100} = S_q p_{cal} = 1.460(10^{-12})(100) = 146(10^{-12}) \text{ C} = 146 \text{ pC}$

$\quad q_{1000} = S_q p_{cal} = 1.460(10^{-12})(1000) = 1460(10^{-12}) \text{ C} = 1460 \text{ pC}$

(d) $\quad v_{cal} = \dfrac{q_{cal}}{C_{cal}} = \dfrac{146(10^{-12})}{1000(10^{-12})} = 0.146 \text{ V}$

$\quad v_{cal} = \dfrac{q_{cal}}{C_{cal}} = \dfrac{1460(10^{-12})}{1000(10^{-12})} = 1.460 \text{ V}$

(e) $\quad v_0 = S_v p_{cal} = 0.005(100) = 0.50 \text{ V}$

$\quad v_0 = S_v p_{cal} = 0.005(1000) = 5.00 \text{ V}$

(f) Units of S_q and S_v would change; otherwise, problem would remain the same.

ENGINEERING MEASUREMENTS by J. W. DALLY, W. F. RILEY, AND K. G. McCONNELL

Exercise 9.27

From Table 9.2: $S_f = 1.0$ mV/lb

From Eqs. (9.47) and (9.50): $S_f^* = \left(\dfrac{V_f}{V_a}\right)\dfrac{S_a}{W_p} = \dfrac{S_f}{1 + M}$

We can measure S_f^* by using Eq. (9.50) for various combinations of head and body masses. Equation (9.47) shows how S_f^* varies with mass ratio M. Thus, for one combination, we obtain S_{f1}^* and Eq. (9.47) allows us to estimate M_1. Now, change the head mass or the body mass a known amount. Recalibrate the hammer to obtain S_{f2}^*. This gives M_2. Knowing M_1, M_2 and the mass added to either the head mass or the body mass, we can solve for the original head mass and the original body mass. Then, S_f^* can be predicted for any new mass combination.

Exercise 9.28

$T = 0.75$ ms

(a) $f_n = \dfrac{1}{T} = \dfrac{1}{0.75(10^{-3})} = 1333$ Hz

$\omega_n = 2\pi f_n = 2\pi(1333) = 8378$ rad/s

(b) For the first peak:

$1 + e^{-d\omega_n t_1} = \dfrac{9.01}{5} = 1.802 \qquad e^{d\omega_n t_1} = 1.247$

$d = \dfrac{\ln 1.247}{\omega_n t_p} = \dfrac{\ln 1.247}{8378(0.375)(10^{-3})} = 0.070 = 7\%$

For the second peak:

$1 + e^{-d\omega_n t_2} = \dfrac{7.59}{5} = 1.518 \qquad e^{d\omega_n t_2} = 1.931$

$d = \dfrac{\ln 1.931}{\omega_n t_p} = \dfrac{\ln 1.247}{8378(1.125)(10^{-3})} = 0.070 = 7\%$

(c) $S_p = \dfrac{\Delta V}{\Delta p} = \dfrac{0.005}{3.0} = 1.667(10^{-3})$ V/psi $= 1.667$ mv/psi

Exercise 9.29

$$q_{max} = S_q a_{max} = 41.8(80) = 3{,}344 \text{ pC}$$

Let $b = 0.418$:
$$q^* = \frac{q_{max}}{0.418} = \frac{3344}{0.418} = 8000 \text{ pC}$$

For $C_f = 1000$ pF:
$$v_o = \frac{8000(10^{-12})}{1000(10^{-12})} = 8.0 \text{ volts} < 10 \text{ volts}$$

$$v^*_{cal} = \frac{S_q a}{C_{cal}} = \frac{41.8(10^{-12})(80)}{1000(10^{-12})} = 3.34 \text{ volts}$$

$$S_v = \frac{S_q}{b\, C_f} = \frac{41.8(10^{-12})}{0.418(1000)(10^{-12})} = 0.100 \text{ V/g} = 100 \text{ mV/g}$$

Any combination of $bC_f \approx 418(10^{-12})$ will work.

Exercise 9.30

Exponential decay is controlled by the charge amplifier ($R_f C_f$).

Damped oscillation is due to the charge amplifier or the recorder amplifier system.

No! The transducer is not part of this simulation.

Exercise 9.31

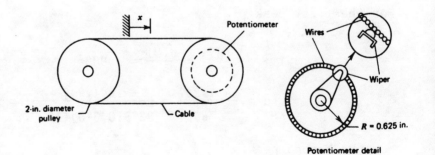

(a) From Eq. (9.59):
$$\mathcal{E} = \frac{1}{1 + 4(R_M/R_p)} = 0.0025$$

Therefore:
$$R_M \cong 100 R_p = 100(20) = 2000 \text{ k}\Omega = 2.0 \text{ M}\Omega$$

(b) $\quad x = R_p \theta = R_p (d_w/R) = 1.0(0.008/0.625) = 0.0128 \text{ in}$

Exercise 9.32

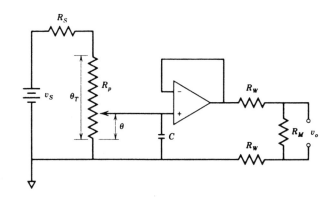

(a) $\quad p = \dfrac{v^2}{R_p} \qquad v = \sqrt{R_p p} = \sqrt{4000(0.02)} = 8.94$ V

(b) With $v_s = 15.0$ V: $\quad iR_s = (15.00 - 8.94) = 6.06$ V

$$i = \dfrac{6.06}{R_s} = \dfrac{8.94}{4000}$$

$\qquad R_s = 6.06(4000)/8.94 = 2711 \; \theta \cong 2.71$ kθ

(c) $\quad v_i = \left(\dfrac{v_s}{R_s + R_p}\right) \dfrac{\theta}{\theta_T} R_p$

$$V_o = i_M R_M = \dfrac{v_i}{R_M + 2R_W} R_M = \left[\dfrac{R_M}{R_M + 2R_W}\right]\left[\dfrac{v_s R_p}{R_s + R_p}\right] \dfrac{\theta}{\theta_T} = S\theta$$

$$S = \left[\dfrac{R_M}{R_M + 2R_W}\right]\left[\dfrac{R_p}{R_s + R_p}\right] \dfrac{v_s}{\theta_T}$$

Exercise 9.33

$S = 10.0$ mV/deg $\quad v_s = 10$ volts $\quad i_M = 20$ mA (max)

$\ell = 400$ ft $\quad R_W = 66.2(0.4) = 26.48\ \Omega$

From Exercise 9.32:

$$S = \left[\frac{R_M}{R_M + 2R_W}\right]\left[\frac{R_p}{R_s + R_p}\right]\frac{v_s}{\theta_T}$$

$$0.01 = \left[\frac{R_M}{R_M + 2(26.5)}\right]\left[\frac{4{,}000}{4{,}000 + R_s}\right]\frac{10}{320}$$

$$\left[\frac{R_M}{R_M + 53}\right]\left[\frac{4{,}000}{4{,}000 + R_s}\right] = 0.320$$

To satisfy the power dissipating capability of the potentiometer:

$$i = \sqrt{p/R_p} = \sqrt{0.02/4000} = 2.24(10^{-3})\ \text{A} = 2.24\ \text{mA (maximum)}$$

For an 8-V drop across R_p with $i_p \cong 2$ mA:

$$R_s = \frac{10 - 8}{i_p} = \frac{10 - 8}{2(10^{-3})} = 1000\ \Omega\ \text{(minimum)}$$

$$\left[\frac{R_M}{R_M + 53}\right]\left[\frac{4{,}000}{4{,}000 + 1000}\right] = 0.320 \qquad R_M = 35.3\ \Omega$$

$$V_{max} = S\theta = 0.01(320) = 3.2\ \text{volts}$$

With $V_{max} = 3.2$ V and $R_M = 35.3\ \Omega$:

$$i_M = v_o/R_M = 3.2/35.3 = 90.7(10^{-3})\ \text{A} = 90.7\ \text{mA (excessive)}$$

With $V_{max} = 3.2$ V and $i_M = 20$ mA:

$$R_M = V_{max}/i_M = 3.2/20(10^{-3}) = 160\ \Omega$$

With $R_M = 160\ \Omega$:

$$\left[\frac{160}{160 + 53}\right]\left[\frac{4{,}000}{4{,}000 + R_s}\right] = 0.320 \qquad R_s = 5390\ \Omega = 5.39\ \text{k}\Omega$$

(a) $R_M = 160\ \Omega$ and $R_s = 5.39$ kΩ.

(b) Add a 10-kΩ trimpot in series with R_p to adjust sensitivity.

(c) Apply known rotations - adjust trimpot to obtain 10 mv/deg.

Exercise 9.34

For the circuit shown Fig. 9.35:

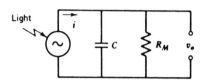

$$i = Ky \qquad i_1 = C\frac{dv_0}{dt} \qquad i_2 = \frac{v_0}{R_M}$$

$$i_1 + i_2 = i$$

$$C\frac{dv_0}{dt} + \frac{v_0}{R_M} = Ky$$

$$R_M C \frac{dv_0}{dt} + v_0 = R_M Ky$$

For the circuit shown Fig. 9.36:

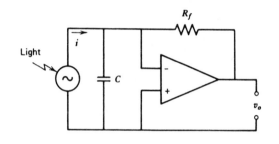

$$i = Ky \qquad v_0 = -Gv_1 \qquad v_1 = -\frac{v_0}{G}$$

$$i_1 = C\frac{dv_1}{dt} = -\frac{C}{G}\frac{dv_0}{dt}$$

$$i_2 = \frac{v_1 - v_0}{R_f} = -\frac{(1+G)v_0}{GR_f}$$

$$i_1 + i_2 = i$$

$$-\frac{C}{G}\frac{dv_0}{dt} - \frac{(1+G)}{G}\frac{v_0}{R_f} = Ky$$

$$\frac{CR_f}{G}\frac{dv_0}{dt} + \frac{1+G}{G}v_0 = -R_f Ky$$

For large G: $\quad v_0 \cong -R_f Ky$ (small time derivative effects).

For the circuit of Fig. 9.35:

$$[1 + jRC\omega]\, v_0 = R_M K y_0$$

↑decay with high freq.

For the circuit of Fig. 9.36 with $G > 10{,}000$ and $R_M = R_f$:

$$\left(1 + j\frac{RC\omega}{G}\right) v_0 = -R_f k y_0$$

↑decay reduced by 10,000 (open loop gain)

Thus, using an op-amp improves high frequency response.

Exercise 9.35

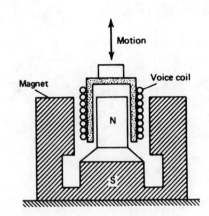

From Eq. (9.60):

$v_T = B\ell V$

From Eq. (1.6):

$S = \dfrac{v_T}{V} = B\ell = 0.060$ V/(m/s)

$\ell = N\pi d = 100(\pi)(0.020) = 6.283$ m

$B = \dfrac{S}{\ell} = \dfrac{0.060}{6.283} = 0.00955$ Wb/m^2

Exercise 9.36

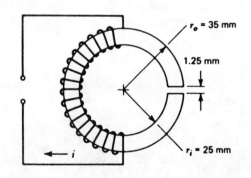

$B = 12.57(10^{-7})\dfrac{Ni}{\ell}$

$\ell = \dfrac{\pi d}{2} = \dfrac{\pi(0.030)}{2}$

$ = 0.0471$ m

$i = \dfrac{B\ell}{12.57(10^{-7})N} = \dfrac{0.10(0.0471)}{12.57(10^{-7})(1000)} = 3.75$ amps

$v_o = B\ell V = 0.10(0.010)(1.0) = 0.001$ V $= 1.0$ mV

Exercise 9.37

From Eq. (9.62)

$$H(\omega) = \frac{R_M S_v}{(R_M + R_T) + jL_T\omega} = \frac{R_M S_v}{R + jL_T\omega}$$

(a) ($R_M = 100\ \Omega$, $R_T = 6.2\ \Omega$, and $L_T = 0.0165$ H):

$$R = R_T + R_M = 106.2$$

$$\frac{R_M}{R} = \frac{100}{106.2} = 0.942\ (> 5\%\text{ error at all frequencies})$$

$$\tan\phi = \frac{L\omega}{R}$$

$$\omega = \frac{R}{L}\tan\phi = \frac{106.2}{0.0165}\tan 10°$$

$$= 1135\text{ rad/s} = 180\text{ Hz}$$

Thus, the 100-Ω instrument is unsatisfactory at any frequency if magnitude errors are to be less than 5 percent.

(b) ($R_M = 1000\ \Omega$, $R_T = 6.2\ \Omega$, and $L_T = 0.0165$ H):

$$R = R_T + R_M = 1006.2\ \Omega$$

Thus:
$$\frac{R_M}{\sqrt{R^2 + (L\omega)^2}} = \frac{1000}{\sqrt{(1006.2)^2 + (0.0165\omega)^2}} = 0.95$$

From which:
$$\omega = 18{,}738\text{ rad/s} = 2982\text{ Hz (maximum frequency)}$$

$$\tan\phi = \frac{L\omega}{R}$$

$$\omega = \frac{R}{L}\tan\phi = \frac{1006.2}{0.0165}\tan 10°$$

$$= 10{,}753\text{ rad/s} = 1710\text{ Hz}$$

Thus, the 1000-Ω instrument is satisfactory for frequencies less than 1710 Hz. The 10-degree phase shift specification controls the upper frequency limit.

Exercise 9.38

$$N = 128\text{ teeth}$$

$$\omega = 2650\text{ rpm} = 44.17\text{ rev/s}$$

$$f = N\omega = 128(44.17) = 5653\text{ pulses/s}$$

Exercise 9.39

$$V = 100 \text{ mm/s} = 0.100 \text{ m/s}$$
$$\lambda = 628 \text{ nm} = 628(10^{-9}) \text{ m}$$

From Eq. (9.64):

$$f_d = \frac{2V}{\lambda} = \frac{2(0.100)}{628(10^{-9})}$$
$$= 318.5(10^3) \text{ Hz} = 318.5 \text{ kHz}$$

Exercise 9.40

For an Endevco 2222C accelerometer (see Table 9.1):

$$S_q^* = 1.4 \text{ pC/g} \qquad S_v = 1.8 \text{ mV/g} \qquad C_t = 470 \text{ pF}$$

For transducer cable: $\qquad C_c = 30 \text{ pF/ft}$

From Eq. (9.22):
$$C = \frac{S_q^*}{S_v} = \frac{1.4(10^{-12})}{1.8(10^{-3})} = 778(10^{-12}) \text{ F} = 778 \text{ pF}$$

$$C_c = C - C_t = 778 - 470 = 308 \text{ pF}$$
$$\ell = \frac{308}{30} = 10.27 \text{ ft} = 123 \text{ in.}$$

For 3-dB cutoff:

$$RC\omega = 1.0$$
$$\omega = \frac{1.0}{100(10^6)(778)(10^{-12})} = 12.85 \text{ rad/s} = 2.05 \text{ Hz}$$

$$C = C_t + C_c = 470 + 30(15) = 920 \text{ pF}$$

$$S_v = \frac{S_q^*}{C} = \frac{1.40(10^{-12})}{920(10^{-12})} = 1.52(10^{-3}) \text{ V/g} = 1.52 \text{ mV/g}$$

For 10% attenuation $RC\omega = 2.06$:

$$\omega = \frac{2.06}{100(10^6)(920)(10^{-12})} = 22.4 \text{ rad/s} = 3.56 \text{ Hz}$$

Exercise 10.1

For the periodic signal shown in Fig. E10.1, the repeat period is $3T_0$.

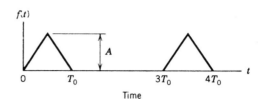

From Eq (10.1):

$$\bar{f} = \frac{1}{3T_0} \int_0^{3T_0} f(t)\, dt = \frac{1}{3T_0} \int_0^{T_0} A\, dt + 0 = \frac{A}{3} \quad \text{(mean)}$$

From Eq. (10.2):

$$\overline{f^2} = \frac{1}{3T_0} \int_0^{3T_0} f^2(t)\, dt = \frac{1}{3T_0} \int_0^{T_0} A^2\, dt = \frac{A^2}{3} \quad \text{(mean square)}$$

From Eq. (10.3):

$$A_{rms} = \sqrt{\overline{f^2}} = \sqrt{\frac{A^2}{3}} = \frac{\sqrt{3}\, A}{3} \quad \text{(root mean square)}$$

Exercise 10.2

For the periodic signal shown in Fig. E10.2, the repeat period is $3T_0$.

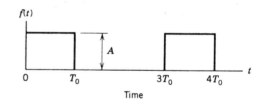

From Eq (10.1):

$$\bar{f} = \frac{1}{3T_0} \int_0^{3T_0} f(t)\, dt = \frac{2}{3T_0} \int_0^{T_0/2} \frac{2At}{T_0}\, dt + 0 = \frac{A}{6} \quad \text{(mean)}$$

From Eq. (10.2):

$$\overline{f^2} = \frac{1}{3T_0} \int_0^{3T_0} f^2(t)\, dt = \frac{2}{3T_0} \int_0^{T_0/2} \frac{4A^2 t^2}{T_0^2}\, dt = \frac{A^2}{9} \quad \text{(mean square)}$$

From Eq. (10.3):

$$A_{rms} = \sqrt{\overline{f^2}} = \sqrt{\frac{A^2}{9}} = \frac{A}{3} \quad \text{(root mean square)}$$

Exercise 10.3

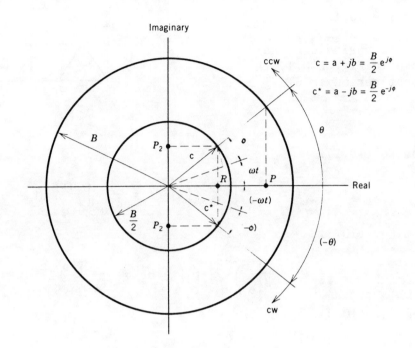

$$f(t) = B \cos(\omega t + \phi) = \left(\frac{B}{2} e^{j\phi}\right) e^{j\omega t} + \left(\frac{B}{2} e^{-j\phi}\right) e^{-j\omega t}$$

$$\cos\theta = \frac{e^{j\theta} + e^{-j\theta}}{2} \qquad \sin\theta = \frac{e^{j\theta} - e^{-j\theta}}{2j}$$

$$\theta = \omega t + \phi \qquad e^{j\theta} = e^{j\phi} e^{j\omega t}$$

$$c = \frac{B}{2} e^{j\phi} \qquad c^* = \frac{B}{2} e^{-j\phi}$$

$$c = \frac{B}{2}(\cos\phi + j\sin\phi) \qquad c^* = \frac{B}{2}(\cos\phi - j\sin\phi)$$

$$= a + jb \qquad = a - jb$$

$$a = \frac{B}{2}\cos\phi \qquad b = \frac{B}{2}\sin\phi$$

$$\tan\phi = \frac{\text{Im}}{\text{Re}} = \frac{b}{a}$$

Exercise 10.4

$$\bar{f} = \frac{1}{T}\int_0^T f(t)\, dt = \frac{1}{T}\int_0^T [D + B \cos(\omega t + \phi)]\, dt$$

$$= \frac{1}{T}\left[Dt + \frac{B \sin(\omega t + \phi)}{\omega}\right]_0^T$$

$$= D + B\left[\frac{\sin(\omega T + \phi) - \sin\phi}{\omega T}\right] \cong D$$

The bracketed term vanishes with increasing time T.

$$A_{rms}^2 = \overline{f^2} = \frac{1}{T}\int_0^T f^2(t)\, dt$$

$$= \frac{1}{T}\int_0^T [D + B\cos(\omega t + \phi)]^2\, dt$$

$$= \frac{1}{T}\int_0^T [D^2 + 2DB\cos(\omega t + \phi) + B^2 \cos^2(\omega t + \phi)]\, dt$$

$$= D^2 + \frac{B^2}{2} + 2DB\left[\frac{\sin(\omega T + \phi) - \sin\phi}{\omega T}\right]$$

$$+ \frac{B^2}{2}\left[\frac{\sin(2\omega T + 2\phi) - \sin 2\phi}{2\omega T}\right]$$

$$\cong D^2 + \frac{B^2}{2}$$

The $\frac{B^2}{2}$ error term starts at 1 for $\omega T = 0$ and decreases very quickly as shown in the following graph.

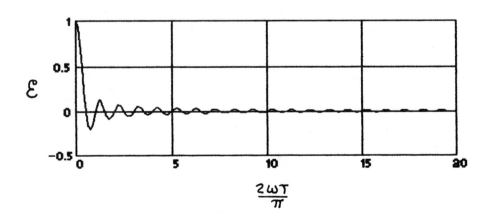

Exercise 10.5

For the periodic signal shown in Fig. E10.1, the repeat period is $3T_0$.

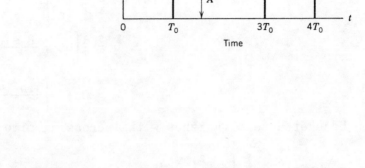

$$\omega_0 = 2\pi/3T_0$$

$f(t) = A \quad 0 \leq t \leq T_0$
$f(t) = 0 \quad T_0 \leq t \leq 3T_0$

From Eq (10.13b):

$$c_p = \frac{1}{T} \int_t^{t+T} f(t)\, e^{-jp\omega_0 t}\, dt$$

Let
$$z = -jp\omega_0 t \qquad dz = -jp\omega_0\, dt$$

$$c_p = \frac{1}{T} \int_0^{T_0} A\, e^{-jp\omega_0 t}\, dt$$

$$= \frac{-A}{jp\omega_0 T} \int_0^{-jp\omega_0 T_0} e^z\, dz$$

$$= \frac{-A}{jp\omega_0 T}\left(e^{-jp\omega_0 T_0} - 1\right)$$

$$= \frac{jA}{3p\omega_0 T_0}\left(\cos p\omega_0 T_0 - 1 - j\sin p\omega_0 T_0\right)$$

$$= \frac{A}{3}\left[\frac{\sin(p\omega_0 T_0)}{(p\omega_0 T_0)}\right] - j\left[\frac{1 - \cos(p\omega_0 T_0)}{(p\omega_0 T_0)}\right]$$

$$= \frac{A}{3}\left[\frac{\sin z}{z} - j\frac{1 - \cos z}{z}\right]$$

$$|c_p| = \frac{A}{3}\left[\frac{\sin^2 z}{z^2} + \frac{1 - 2\cos z + \cos^2 z}{z^2}\right]^{1/2}$$

$$= \frac{A}{3}\left[\frac{2(1 - \cos z)}{z^2}\right]^{1/2}$$

$$\tan \phi_p = \left[\frac{1 - \cos z}{-\sin z}\right]$$

Exercise 10.5 (Continued)

p	z	$3c_p/A$	ϕ
0	0	1.0	0
1	$2\pi/3$	0.827	-60
2	$4\pi/3$	0.413	-120
3	π	0	0
4	$8\pi/3$	0.207	-60
5	$10\pi/3$	0.165	-120
6	2π	0	0

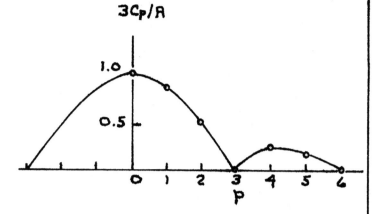

Note that every third term is missing.
Same plot as Fig. 10.6 when β = 3.0.

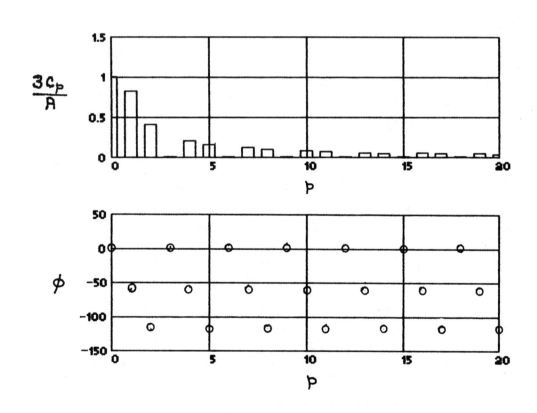

Exercise 10.6

For the periodic signal shown in Fig. E10.2, the repeat period is $3T_0$.

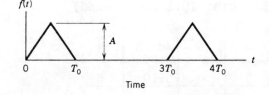

$$\omega_0 = 2\pi/3T_0$$

$f_1(t) = 2At/T_0 \qquad 0 \le t \le T_0/2$
$f_2(t) = 2A(1 - t/T_0) \qquad T_0/2 \le t \le T_0$
$f_3(t) = 0 \qquad T_0 \le t \le 3T_0$

From Eq (10.13b):
$$c_p = \frac{1}{T}\int_t^{t+T} f(t)\, e^{-jp\omega_0 t}\, dt$$

Let $\quad \lambda = jp\omega_0 t \qquad d\lambda = jp\omega_0\, dt \qquad z = jp\omega_0 T_0 = jp(2\pi/3)$

$$c_p = \frac{1}{T}\int_0^{T_0/2} f_1(t) e^{-jp\omega_0 t}\, dt + \frac{1}{T}\int_{T_0/2}^{T_0} f_2(t) e^{-jp\omega_0 t}\, dt + \frac{1}{T}\int_{T_0}^{3T_0} (0) e^{-jp\omega_0 t}\, dt$$

$$= \frac{2A}{3z}\left[\int_0^{z/2} \frac{\lambda}{z} e^{-\lambda}\, d\lambda + \int_{z/2}^{z} \left(1 - \frac{\lambda}{z}\right) e^{-\lambda}\, d\lambda + 0\right]$$

$$= \frac{2A}{3z}\left[\frac{1}{z}\left(1 - \left(\frac{z}{2} + 1\right)e^{-z/2}\right) + \frac{(1+z)}{z}e^{-z} - \frac{(1+z/2)}{z}e^{-z/2} + e^z - e^{-z}\right]$$

$$= \frac{-2A}{3z^2}\left[2e^{-z/2} - 1 - e^{-z}\right]$$

$$= \frac{3A}{2\pi^2 p^2}\left\{2\left[\cos\left(\frac{p\pi}{3}\right) - j\sin\left(\frac{p\pi}{3}\right)\right] - 1 - \cos\left(\frac{p2\pi}{3}\right) + j\sin\left(\frac{p2\pi}{3}\right)\right\}$$

$$= \frac{3A}{2\pi^2 p^2}\left\{\left[2\cos\left(\frac{p\pi}{3}\right) - 1 - \cos\left(\frac{p2\pi}{3}\right)\right] + j\left[\sin\left(\frac{p2\pi}{3}\right) - 2\sin\left(\frac{p\pi}{3}\right)\right]\right\}$$

Note in the figure below that the zero values occur at multiples of 6 instead of 3 and drop off faster than those for a square pulse.

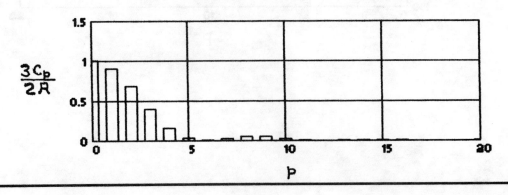

Exercise 10.7

For the periodic signal shown in Fig. E10.1, the repeat period is $3T_0$.

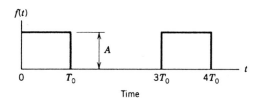

From Exercise 10.1:

$$\bar{f} = \frac{1}{3T_0} \int_0^{3T_0} f(t)\, dt = \frac{A}{3} \quad \text{(mean)}$$

$$\overline{f^2} = \frac{1}{3T_0} \int_0^{3T_0} f^2(t)\, dt = \frac{A^2}{3} \quad \text{(mean square)}$$

$$A_{rms}^2 = \overline{f^2} = \frac{A^2}{3} = 0.333A^2 \quad \text{(root mean square)}$$

From Eq. (10.16):

$$A_{rms}^2 = c_0^2 + 2\sum_{p=1}^{\infty} |c_p|^2$$

From Exercise 10.5:

$$|c_p| = \frac{A}{3}\left[\frac{2(1 - \cos z)}{z^2}\right]^{1/2} \qquad z = p\left(\frac{2\pi}{3}\right)$$

Using 20 terms yields: $A_{rms}^2 = 0.329A^2$

Using 40 terms yields: $A_{rms}^2 = 0.331A^2$

Using 100 terms yields: $A_{rms}^2 = 0.332A^2$

Exercise 10.8

For the periodic signal shown in Fig. E10.2, the repeat period is $3T_0$.

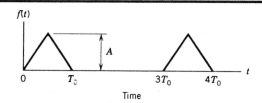

From Exercise 10.2:

$$\bar{f} = \frac{1}{3T_0} \int_0^{3T_0} f(t)\, dt = \frac{A}{6} \quad \text{(mean)}$$

$$\overline{f^2} = \frac{1}{3T_0} \int_0^{3T_0} f^2(t)\, dt = \frac{A^2}{9} \quad \text{(mean square)}$$

$$A_{rms}^2 = \overline{f^2} = \frac{A^2}{9} = 0.111A^2 \quad \text{(root mean square)}$$

From Eq. (10.16):

$$A_{rms}^2 = c_0^2 + 2\sum_{p=1}^{\infty} |c_p|^2$$

From Exercise 10.6:

$$c_p = \frac{-2A}{3z^2}\left[2e^{-z/2} - 1 - e^{-z}\right] \qquad z = jp\left(\frac{2\pi}{3}\right)$$

Using 20 terms yields: $A_{rms}^2 = 0.111A^2$

Exercise 10.9

For the half-sine pulse shown in Fig. E10.9:

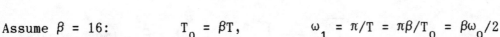

$f(t) = A \sin \omega_1 t \qquad 0 \leq t \leq T$

$f(t) = 0 \qquad t > T$

Assume $\beta = 16$: $\qquad T_0 = \beta T, \qquad \omega_1 = \pi/T = \pi\beta/T_0 = \beta\omega_0/2$

Fourier series:

$$c_p = \frac{1}{T_0} \int_0^{T_0} f(t) e^{-jp\omega_0 t} \, dt = \frac{A}{T_0} \int_0^T \sin(\omega_1 t) e^{-jp\omega_0 t} \, dt$$

$$= \frac{A}{T_0} \left[\frac{e^{-jp\omega_0 t}[jp\omega_0 \sin \omega_1 t - \omega \cos \omega_1 t]}{(jp\omega_0)^2 + \omega_1^2} \right]_0^T$$

$$= \frac{A\omega_1}{T_0} \left[\frac{1 + \cos(p\omega_0 T) - j \sin(p\omega_0 T)}{\omega_1^2 - p^2 \omega_0^2} \right]$$

Fourier Transform:

$$C(\omega) = \int_{-\infty}^{+\infty} f(t) e^{-j\omega t} \, dt = A \int_0^T \sin(\omega_1 t) e^{-j\omega t} \, dt$$

$$= A \left[\frac{e^{-j\omega t}[-j\omega \sin(\omega_1 t) - \omega_1 \cos(\omega_1 t)]}{\omega_1^2 - \omega^2} \right]_0^T$$

$$= A\omega_1 \left[\frac{1 + \cos(\omega t) - j \sin(\omega t)}{\omega_1^2 - \omega^2} \right]$$

Comparing the equations for c_p and $C(\omega)$ shows:

$$C(\omega) = c_p T_0 \qquad \text{and} \qquad \omega = p\omega_0$$

Also, note that in the limit as $\omega \to \omega_1$:

$$|C(\omega)| = \frac{A}{2\omega_1} \qquad \text{and} \qquad |c_p T_0| = \frac{A}{2\omega_1}.$$

For $\beta < 10$, the two solutions plot as one. Both magnitudes have the form:

$$|c_p T_0| = A\omega_1 \frac{\sqrt{2(1 + \cos \omega t)}}{\omega_1^2 - \omega^2} = A\omega_1 \frac{\sqrt{2(1 + \cos p\omega_0 T)}}{\omega_1^2 - p^2 \omega_0^2}$$

Exercise 10.10

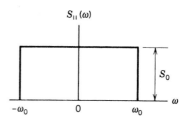

From Eq. (10.24):

$$\phi_{11}(0) = \frac{1}{2\pi} \int_{-\infty}^{+\infty} S_{11}(\omega)\, d\omega$$

$$= \frac{S_0}{2\pi} \int_{-\omega_0}^{+\omega_0} d\omega = \frac{S_0(2\omega_0)}{2\pi} = \frac{S_0 \omega_0}{\pi}$$

$$\phi_{11}(\tau) = \frac{S_0}{2\pi} \int_{-\omega_0}^{+\omega_0} e^{j\omega\tau}\, d\omega$$

$$= \frac{S_0}{j2\pi\tau} \left[e^{j\omega\tau} \right]_{-\omega_0}^{+\omega_0}$$

$$= \frac{S_0}{j2\pi\tau} \left(e^{j\omega_0\tau} - e^{-j\omega_0\tau} \right) = \frac{S_0 \omega_0}{\pi} \frac{\sin \omega_0 \tau}{\omega_0 \tau} = \frac{S_0 \omega_0}{\pi} \operatorname{sinc}(\omega_0 t)$$

Auto correlation is a sinc function.

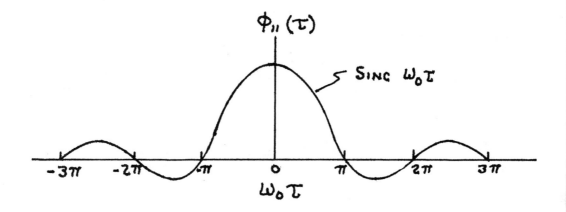

Exercise 10.11

$$k_1 = k \qquad k_2 = 3k \qquad k_3 = 0 \qquad m_1 = 2m \qquad m_2 = 3m$$

From Eq. (10.27):
$$2m\ddot{x}_1 + 4kx_1 - 3kx_2 = 0$$
$$3m\ddot{x}_2 - 3kx_1 + 3kx_2 = 0$$

Substitute: $\quad x_1 = A_1 e^{j\omega t} \quad$ and $\quad x_2 = A_2 e^{j\omega t}$

$$(4k - 2m\omega^2)A_1 - 3kA_2 = 0$$
$$-3kA_1 + (3k - 3m\omega^2)A_2 = 0$$

Therefore:
$$\frac{A_1}{A_2} = \frac{3k}{4k - 2m\omega^2} = \frac{3(k - m\omega^2)}{3k} \qquad \text{Eq. (1)}$$

Which yields:
$$9k^2 = 6(2k^2 - 3km\omega^2 + m^2\omega^4)$$

or:
$$\omega^4 - 3\left(\frac{k}{m}\right)\omega^2 + 0.5\left(\frac{k}{m}\right)^2 = 0$$

Which has the solution:
$$\omega^2 = 0.177\left(\frac{k}{m}\right) \text{ and } 2.82\left(\frac{k}{m}\right)$$

Thus: $\quad \omega_1 = 0.421\sqrt{k/m} \quad$ and $\quad \omega_2 = 1.680\sqrt{k/m} \quad$ (natural frequency)

Mode shapes from Eq. (1):

$$\frac{A_1^{(1)}}{A_2} = \frac{3k}{4k - 2m\omega_1^2} = \frac{3k}{4k - 2m(0.177)k/m} = \frac{3}{4 - 0.354} = 0.823$$

$$\frac{A_1^{(2)}}{A_2} = \frac{3k}{4k - 2m\omega_2^2} = \frac{3k}{4k - 2m(2.82)k/m} = \frac{3}{4 - 2(2.82)} = -1.829$$

Mode shapes:
$$\{u\}^1 = \begin{Bmatrix} 0.823 \\ 1.0 \end{Bmatrix} \qquad \{u\}^2 = \begin{Bmatrix} -1.829 \\ 1.0 \end{Bmatrix}$$

Exercise 10.12

$H_{mn}(\omega)$ is the frequency response function at location m due to excitation at location n. We see from Eq. (10.45) that these FRF's can be obtained through either direct solutions that use dynamic stiffness $D_{ij}(\omega)$ and the system determinant Δ or the modal vector components u_{ij} and the modal dynamic stiffness. Eq. (10.45) shows that the same responses should be obtained from either method.

Exercise 10.13

$$V_k(x) = \sin \frac{k\pi x}{\ell}$$

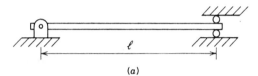

From Eq. (10.53):

$$Q_k = \int_0^\ell P(x)\, V_k(x)\, dx$$

$$= P_0 \int_0^\ell \sin \frac{k\pi x}{\ell}\, dx$$

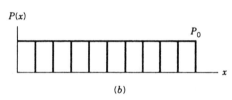

Let $z = \dfrac{k\pi x}{\ell}$

$$Q_k = \frac{P_0 \ell}{k\pi} \int_0^{k\pi} \sin z\, dz = \frac{P_0 \ell}{k\pi}\left[-\cos z\right]_0^{k\pi} = \frac{P_0 \ell}{k\pi}(1 - \cos k\pi)$$

$$Q_k = \frac{2P_0 \ell}{k\pi} \quad \text{for k odd} \qquad Q_k = 0 \quad \text{for k even}$$

Implication is that only odd modes of vibration are excited.

Exercise 10.14

With the accelerometer attached at the midpoint:

From Eq. (10.55):
$$y(b,t) = \sum_{k=1}^{\infty} V_k(b)\, q_k(t) = \sum_{k=1}^{\infty} \frac{V_k(b)\, Q_k}{D_k} e^{j\omega t}$$

With $b = \ell/2$:
$$V_k(b) = \sin \frac{k\pi}{2} = 0 \quad \text{for k even}$$
$$= \pm 1 \quad \text{for k odd}$$

Therefore, only odd natural frequencies are observable. Exercise 10.13 indicated that only odd natural frequencies are excited. Thus, only odd natural frequencies are excited and observed.

Exercise 10.15

$$V_k(x) = \sin \frac{k\pi x}{\ell}$$

$$P(x) = -P_0 + \frac{2P_0}{\ell} x$$

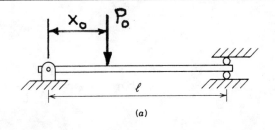

From Eq. (10.53):

$$Q_k = \int_0^\ell P(x) V_k(x) dx$$

$$= \frac{P_0}{\ell} \int_0^\ell (2x - \ell) \sin \frac{k\pi x}{\ell} dx = -\frac{P_0 \ell}{k\pi}(1 + \cos(k\pi))$$

$$Q_k = -\frac{2P_0 \ell}{k\pi} \quad \text{for k even} \qquad Q_k = 0 \quad \text{for k odd}$$

This excitation force excites only even modes of vibration.

Exercise 10.16

$$V_k(x) = \sin \frac{k\pi x}{\ell}$$

$$P(x) = P_0 \, \delta(x - x_0)$$

From Eq. (10.53):

$$Q_k = \int_0^\ell P(x) V_k(x) dx = P_0 \int_0^\ell \delta(x - x_0) \sin \frac{k\pi x}{\ell} dx$$

Since:

$$\int_0^\ell \delta(x - x_0) dx = \int_{x_0^-}^{x_0^+} \delta(x - x_0) dx = 1.0$$

only the value $V_k(x_0)$ comes through integration. Therefore:

$$Q_k = P_0 \sin \frac{k\pi x_0}{\ell}$$

When $x_0 = \ell/3$. $\qquad Q_k = P_0 \sin \frac{k\pi}{3}$

Whenever k is an integer that is divisible by 3 the value of Q_k is zero.

ENGINEERING MEASUREMENTS by J. W. DALLY, W. F. RILEY, AND K. G. McCONNELL

Exercise 10.17

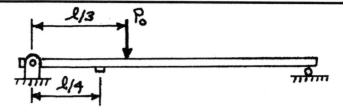

From Exercise 10.16:

With $x_0 = \ell/3$: $\quad Q_k = P_0 \sin \dfrac{k\pi}{3}$ (Modes 3, 6, 9, $\cdots$ eliminated).

With $x_t = \ell/4$: $\quad V_k(\ell/4) = \sin \dfrac{k\pi}{4}$ (Modes 4, 8, 12, $\cdots$ eliminated).

$$\text{excitation eliminates}$$
$$\downarrow \quad \downarrow \quad \downarrow \quad \downarrow \quad \downarrow$$
Thus, 1, 2, 3, 4, 5, 6, 7, 8, 9, 10, 11, 12, 13, 14, 15, 16
$$\uparrow \quad \uparrow \quad \uparrow \quad \uparrow$$
$$\text{transducer location eliminates}$$

We should find natural frequencies 1, 2, 5, 6, 7, 10, 11, 13, 14, 16, etc.

Exercise 10.18

The response to the half-sine impulse is shown in the computer plot below:

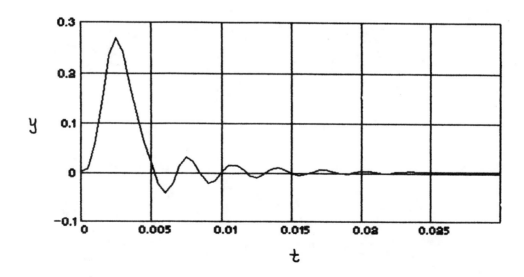

Exercise 10.19

From Eqs (10.63):
$$h(t) = \frac{1}{2\pi} \int_{-\infty}^{+\infty} H(\omega) e^{j\omega t} d\omega \tag{1}$$

$$H(\omega) = \int_{-\infty}^{+\infty} h(t) e^{-j\omega t} dt \tag{2}$$

For:
$$h(t) = \frac{1}{m\omega_n} \sin \omega_n t \quad t > 0 \\ = 0 \quad t < 0 \quad \Big\} \text{ impulse response}$$

Eq. (2) becomes:

$$H(\omega) = \frac{1}{m\omega_n} \int_0^\infty \sin(\omega_n t) e^{-j\omega t} dt$$

$$= \frac{1}{m\omega_n} \left[\frac{e^{-j\omega t}(-j\omega \sin(\omega_n t) - \omega_n \cos(\omega_n t))}{\omega_n^2 + (j\omega)^2} \right]_0^\infty$$

$$= \frac{\omega_n}{m\omega_n(\omega_n^2 - \omega^2)} = \frac{1}{m(\omega_n^2 - \omega^2)} = \frac{1}{k - m\omega^2}$$

Evaluation of Eq. (1) requires an elaborate integral table.

Exercise 10.20

From Eq. (10.69):
$$F_1(\omega) = \frac{1}{2\pi} \int_{-\infty}^{+\infty} F_2(\nu) F_3(\omega - \nu) d\nu$$

$$= \int_{-\infty}^{+\infty} f_2(t) f_3(t) e^{-j\omega t} dt \tag{1}$$

Let
$$f_2(t) = \frac{1}{2\pi} \int_{-\infty}^{+\infty} F_2(\nu) e^{j\nu t} d\nu$$

Substituting into Eq. (1) and interchanging position of $e^{j\nu t}$ yields:

$$F_1(\omega) = \frac{1}{2\pi} \int_{-\infty}^{+\infty} F_2(\nu) \left\{ \int_{-\infty}^{+\infty} f_3(t) e^{-j(\omega-\nu)t} dt \right\} d\nu$$

But:
$$F_3(\omega - \nu) = \int_{-\infty}^{+\infty} f_3(t) e^{-(\omega-\nu)t} dt$$

Thus:
$$F_2(\omega) = \frac{1}{2\pi} \int_{-\infty}^{+\infty} F_2(\nu) F_3(\omega - \nu) d\nu$$

Convolution in the frequency domain.

Exercise 10.21

On the sketch at the right:

$$f_b = 0.4 f_s$$

For a 14 bit A/D converter:

$$N_{dB} = 20 \log(2^{14}) = 84 \text{ dB}$$

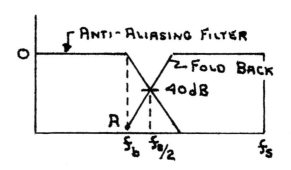

To be down at least 80 dB at $0.4 f_s$ (point A), the attenuation must be achieved over a very small Δf; i.e. $\approx 0.2 f_s$ from $0.6 f_s$ to $0.4 f_s$ (see the foldback curve on the sketch). For a slope of 120 dB/octave (log scale)

f/f_s	f/f_0	ΔdB
0.4f	1.0	0
0.5f	1.25	38.6
0.6f	1.50	70.2
0.7f	1.75	97
0.8f	2.00	120

$$S = \frac{\Delta dB}{\log(f/f_0)} = \frac{120}{\log(2)} = 398.6$$

$$\Delta dB = 398.6 \log(f/f_0) \qquad f_0 = 0.4 f_s$$

would like to be down 40 dB by the Nyquist frequency

Exercise 10.22

From Eq. (10.74):
$$B_{rms}^2 = 2 \sum |C_p|^2$$

$$C_p^2 = 2 \left\{ \frac{\text{sinc}^2[(p+N)\pi]}{(1 - N/p)} + \frac{\text{sinc}^2[(p-N)\pi]}{(1 + N/p)} \right\} \quad \begin{array}{l} N = 10.25 \\ N = 10.5 \end{array} \quad 7 < p < 14$$

P	N = 10.25 $2c_p^2$	N = 10.50 $2c_p^2$	P	N = 10.25 $2c_p^2$	N = 10.50 $2c_p^2$
7	0.003	0.005	11	0.097	0.424
8	0.008	0.012	12	0.019	0.051
9	0.028	0.038	13	0.008	0.020
10	0.791	0.386	14	0.006	0.011

$$2\Sigma c_p^2 = 0.959 \qquad 0.948$$
$$B_{rms} = 0.979 \qquad 0.974$$

Thus: B_{rms} is within 2.5% of the original value of unity.

ENGINEERING MEASUREMENTS by J. W. DALLY, W. F. RILEY, AND K. G. McCONNELL

Exercise 10.23

From Eq. (10.7): $\quad c = \dfrac{B}{2} e^{j\phi} \quad$ for any frequency.

From Eq. (10.8): $\quad c^* = \dfrac{B}{2} e^{-j\phi} \quad$ complex conjugate.

Thus: $\quad cc^* = \dfrac{B^2}{4} e^{j\phi} e^{-j\phi} = \dfrac{B^2}{4}$

$$B = 2\sqrt{cc^*} = 2|c| \qquad (1)$$

$$B_{rms} = \dfrac{B}{\sqrt{2}} = \dfrac{2\sqrt{cc^*}}{\sqrt{2}} = \sqrt{2cc^*} \qquad (2)$$

Eq. (1) and (2) hold for any frequency component (obtained in a frequency analyzer). These equations allow us to manipulate any frequency component to obtain single or double sided peak component displays, RMS single sided displays, mean square spectral displays, and spectral density displays.

Exercise 10.24

From Eq. (10.85): $\quad \epsilon_r = \dfrac{1}{2\sqrt{N_r}}$

With N_r = 75 and 150: $\quad \epsilon_{r1} = \dfrac{1}{2\sqrt{75}} = 0.0577$

$$\epsilon_{r2} = \dfrac{1}{2\sqrt{150}} = 0.0408$$

Therefore: $\quad \Delta\epsilon_r = \dfrac{0.0577 - 0.0408}{0.0577}(100) = 29.3\ \%$

Exercise 10.25

As shown in Fig. 10.23, the Kaiser-Bessel window function has the required frequency resolution and the amplitude drop off to detect the 40-dB signal difference. The high side lobes of the Hanning window would not permit detection of the 40-dB signal difference.

Exercise 10.26

These windows attenuate the signal unless it is centered in the time window. Thus, the corresponding frequency spectra is also attenuated.

Exercise 10.27

The average value for m readings is:

$$\bar{x}_m = \frac{1}{m} \sum_{k=1}^{m} x_k \qquad (1)$$

The average value for m + 1 readings is:

$$\bar{x}_{m+1} = \frac{1}{m+1}\left[\sum_{k=1}^{m} x_k + x_{m+1}\right]$$

But:

$$\sum_{k=1}^{m} x_k = m \bar{x}_m$$

Therefore:

$$\bar{x}_{m+1} = \frac{m}{m+1} \bar{x}_m + \left[\frac{x_m + 1}{m+1}\right]$$

Let $m = r - 1$:

$$\bar{x}_r = \left(\frac{r-1}{r}\right) \bar{x}_{(r-1)} + \frac{x_r}{r}$$

This equation has the same form as Eq. (10.86) when x_r is replaced by $|D_p(k)|^2$.

Exercise 10.28

From Eqs. 10.88:

$$S_{zx} = S_0 \sin \phi \cos \theta$$

$$S_{zy} = S_0 \sin \phi \sin \theta$$

$$S_{zz} = S_0 \cos \phi \qquad \text{(primary)}$$

Max. cross axis:

$$\frac{S_0 \sin \phi}{S_0 \cos \phi} = 0.05$$

$$\phi = \tan^{-1} 0.05 = \pm 2.86°$$

Exercise 10.29

From Eq. (10.87):

$$v_x = S_{xx}a_x + S_{xy}a_y + S_{xz}a_z \qquad x \text{ - primary sensing axis}$$

$$v_y = S_{yx}a_x + S_{yy}a_y + S_{yz}a_z \qquad y \text{ - primary sensing axis}$$

$$v_z = S_{zx}a_x + S_{zy}a_y + S_{zz}a_z \qquad z \text{ - primary sensing axis}$$

Divide each equation by its primary-axis sensitivity (S_{xx}, S_{yy}, S_{zz}) and let $b_i = v_i/S_{ii}$ be the apparent acceleration and $\varepsilon_{ip} = S_{ip}/S_{ii}$ be the cross-axis sensitivity. Then,

$$\varepsilon_{xx}a_x + \varepsilon_{xy}a_y + \varepsilon_{xz}a_z = b_x$$

$$\varepsilon_{yx}a_x + \varepsilon_{yy}a_y + \varepsilon_{yz}a_z = b_y$$

$$\varepsilon_{zx}a_x + \varepsilon_{zy}a_y + \varepsilon_{zz}a_z = b_z$$

Exercise 10.30

Given that: $\qquad \varepsilon_{ij} \approx 0.05$

From Eq. (10.92): $\qquad [\varepsilon]\{a\} = \{b\}$

From Eq. (10.93): $\qquad \{a\} = [C]\{b\}$

$$c_{11} = \frac{1 - \varepsilon_{23}\varepsilon_{32}}{\Delta}$$

$$\Delta = 1 - \underbrace{\varepsilon_{23}\varepsilon_{32} - \varepsilon_{12}\varepsilon_{21} - \varepsilon_{13}\varepsilon_{31}}_{\text{2nd order}} + \underbrace{\varepsilon_{12}\varepsilon_{23}\varepsilon_{31} + \varepsilon_{13}\varepsilon_{21}\varepsilon_{32}}_{\text{3rd order}}$$

$$= 1 \pm 3(0.0025) = 1 \pm 0.0075 \quad \text{(With 3rd order terms neglected)}.$$

$$c_{11} = \frac{1 \pm 0.0025}{1 \pm 0.0075} = \frac{0.9975}{1.0075} = 0.990 \qquad \longrightarrow \qquad \frac{1.0025}{0.0025} = 1.01$$

Exercise 10.31

$$C_{ii} \approx 1.0 \approx 1 - \epsilon_{jk}\epsilon_{kj}$$

$$C_{ij} \approx -\frac{\epsilon_{ij} - \epsilon_{ik}\epsilon_{jk}}{\Delta} \approx -\epsilon_{ij}$$

Thus,
$$a_x = b_x - \epsilon_{xy}b_y - \epsilon_{xz}b_z$$

$$a_y = -\epsilon_{yx}b_x + b_y - \epsilon_{yz}b_z$$

$$a_z = -\epsilon_{zx}b_x - \epsilon_{zy}b_y + b_z$$

Exercise 10.32

From Eq. (9.47):
$$S_f^* = \frac{S_f m_2}{m_1 + m_2} = \frac{S_f}{1 + (m_1/m_2)}$$

The effect of the hand is to increase mass m_2 to $2m_2$.

Thus:
$$S_f^* = \frac{2S_f m_2}{m_1 + 2m_2} = \frac{S_f}{1 + (m_1/2m_2)}$$

Thus, the sensitivity increases from:

$$\frac{S_f}{1 + (m_1/m_2)} \quad \text{to} \quad \frac{S_f}{1 + (m_1/2m_2)}$$

Note also that the natural frequency $\omega_n = \sqrt{k/m_e}$ is affected. Since the effective mass m_e increases, the natural frequency decreases.

Exercise 10.33

From Eq. (10.101):
$$v_f = S_f H_f(\omega)\left[1 - m_1 H_{11}(\omega)\,\omega^2\right] F_1$$

From Eq. (h):
$$m_s \ddot{x}_1 + c_s \dot{x}_1 + k_s x_1 = F_1(t)$$

Thus:
$$H_{11}(\omega) = \frac{1}{k_s - m_s \omega^2 + jc_s \omega} = \frac{x_1}{F_1}$$

For an approximate voltage $\hat{v}_f = S_f H_f F_1$:

$$\frac{\hat{v}_f}{v_f} = \frac{1}{1 - m_1 H_{11}(\omega)\,\omega^2} = \frac{k_s - m_s \omega^2 + jc_s \omega}{k_s - (m_s + m_1)\omega^2 + jc_s \omega}$$

$$= \frac{1 - r^2 + j2d_s r}{1 - (1+n)r^2 + j2d_s r}$$

where $n = m_1/m_s$

Exercise 11.1

Temperature is an abstract quantity that is defined in terms of the thermodynamic activity of materials with changes in temperature. The 17 fixed points, defined in the International Temperature Scale, provide reference points for calibrating temperature sensors. Most of these points correspond to the equilibrium state during a phase transition of a particular material.

Exercise 11.2

A bath of finely crushed ice in water at sea level represents a media at $T = 0.01°C$. The sensor is inserted into this bath and maintained until it has achieved thermal equilibrium. The reading of the sensor is recorded at this temperature, and compared to its previous calibration record.

A solder pot filled with pure tin is heated until it is melted. The sensor is inserted into the center of the molten bath and maintained there until it is in thermal equilibrium. Power to the solder pot is cut off and the temperature of the tin drops with time until the freezing process begins. During freezing, the temperature of the tin remains constant with time at $T = 231.928°C$. The sensor output is recorded during this period and compared to its previous calibration.

Exercise 11.3

Above the freezing point of silver (Ag) at $961.78°C$, the temperature of a calibration body is defined using certified optical pyrometers to measure radiation and Planck's law to relate this radiation to the temperature.

Exercise 11.4

Stainless steel: $h_1 = 0.5$ mm $\alpha_1 = 17.3(10^{-6})/°C$ $E_1 = 193$ GPa

Invar: $h_2 = 1.0$ mm $\alpha_2 = 1.1(10^{-6})/°C$ $E_2 = 145$ GPa

From Eq. (11.4):

$$\rho = \frac{\left[3\left(1 + r_h\right)^2 + \left(1 + r_h r_E\right)\left(r_E^2 - \frac{1}{r_h r_E}\right)\right] h}{6(\alpha_1 - \alpha_2)(1 + r_h)^2 \Delta T}$$

For:
$$r_h = h_2/h_1 = 1.0/0.5 = 2.0$$
$$r_E = E_2/E_1 = 145/193 = 0.75$$
$$h = h_1 + h_2 = 1.5$$

$$\rho = \frac{\left\{3(1 + 2)^2 + \left[1 + 2(0.75)\right]\left[(0.75)^2 - \frac{1}{2(0.75)}\right]\right\}(1.5)}{6(17.3 - 1.1)(10^{-6})(1 + 2)^2 \Delta T} = \frac{45{,}850}{\Delta T}$$

(a) For $\Delta T = 120°C$ (216°F): $\rho = 382$ mm (15.04 in.)

(b) For $\Delta T = 70°F$ (38.9°C): $\rho = 1179$ mm (46.4 in.)

(c) For $\Delta T = 230°C$ (414°F): $\rho = 199$ mm (7.83 in.)

(d) For $\Delta T = -80°F$ (-44.4°C): $\rho = -1033$ mm (-40.7 in.)

Exercise 11.5

From Eq. (11.5):
$$R = R_0(1 + \gamma_1 T + \gamma_2 T^2 + \gamma_3 T^3)$$

For a platinum RTD:

Temp	-200	400	1000
R/R_0	0.25	2.42	4.33

$$\gamma_1(-200) + \gamma_2(-200)^2 + \gamma_3(-200)^3 = -0.75$$
$$\gamma_1(400) + \gamma_2(400)^2 + \gamma_3(400)^3 = 1.42$$
$$\gamma_1(1000) + \gamma_2(1000)^2 + \gamma_3(1000)^3 = 3.33$$

Solving yields:
$$\gamma_1 = 3.686(10^{-3})/(°C)$$
$$\gamma_2 = -0.329(10^{-6})/(°C)^2$$
$$\gamma_3 = -0.027(10^{-9})/(°C)^3$$

Exercise 11.6

From Eq. (11.5):
$$R = R_0(1 + \gamma_1 T + \gamma_2 T^2 + \gamma_2 T^3)$$

For a copper RTD:

Temp	-200	400	1000
R/R_0	0	2.83	6.00

$$\gamma_1(-200) + \gamma_2(-200)^2 + \gamma_3(-200)^3 = -1.00$$
$$\gamma_1(400) + \gamma_2(400)^2 + \gamma_3(400)^3 = 1.83$$
$$\gamma_1(1000) + \gamma_2(1000)^2 + \gamma_3(1000)^3 = 5.00$$

Solving yields:
$$\gamma_1 = 4.763(10^{-3})/(°C)$$
$$\gamma_2 = -1.132(10^{-6})/(°C)^2$$
$$\gamma_3 = 1.369(10^{-9})/(°C)^3$$

Exercise 11.7

From Eq. (11.5): $$R = R_0(1 + \gamma_1 T + \gamma_2 T^2 + \gamma_2 T^3)$$

For a nickel RTD:

Temp	100	500	1000
R/R_0	1.58	5.34	7.50

$$\gamma_1(100) + \gamma_2(100)^2 + \gamma_3(100)^3 = 0.58$$
$$\gamma_1(500) + \gamma_2(500)^2 + \gamma_3(500)^3 = 4.34$$
$$\gamma_1(1000) + \gamma_2(1000)^2 + \gamma_3(1000)^3 = 6.50$$

Solving yields:

$$\gamma_1 = 4.4375(10^{-3})/(°C)$$
$$\gamma_2 = 14.91(10^{-6})/(°C)^2$$
$$\gamma_3 = -12.85(10^{-9})/(°C)^3$$

Exercise 11.8

From Eq. (11.5): $$R = R_0(1 + \gamma_1 T + \gamma_2 T^2 + \gamma_3 T^3)$$

For a platinum RTD:

Temp	100	400	700
R/R_0	1.33	2.42	3.50

$$\gamma_1(100) + \gamma_2(100)^2 + \gamma_3(100)^3 = 0.33$$
$$\gamma_1(400) + \gamma_2(400)^2 + \gamma_3(400)^3 = 1.42$$
$$\gamma_1(700) + \gamma_2(700)^2 + \gamma_3(700)^3 = 2.50$$

Solving yields:

$$\gamma_1 = 3.166(10^{-3})/(°C)$$
$$\gamma_2 = 1.467(10^{-6})/(°C)^2$$
$$\gamma_3 = -1.27(10^{-9})/(°C)^3$$

Exercise 11.9

From Eq. (11.5):
$$R = R_0 (1 + \gamma_1 T + \gamma_2 T^2 + \gamma_2 T^3)$$

For a copper RTD:

Temp	-100	200	500
R/R_0	0.5	1.95	3.33

$$\gamma_1(-100) + \gamma_2(-100)^2 + \gamma_3(-100)^3 = -0.5$$
$$\gamma_1(200) + \gamma_2(200)^2 + \gamma_3(200)^3 = 0.95$$
$$\gamma_1(500) + \gamma_2(500)^2 + \gamma_3(500)^3 = 2.33$$

Solving yields:

$$\gamma_1 = 22.04(10^{-3})/(°C)$$
$$\gamma_2 = -10.72(10^{-6})/(°C)^2$$
$$\gamma_3 = -48.08(10^{-9})/(°C)^3$$

Exercise 11.10

From Eq. (11.5):
$$R = R_0 (1 + \gamma_1 T + \gamma_2 T^2 + \gamma_2 T^3)$$

For a nickel RTD:

Temp	100	200	400
R/R_0	1.5	2.58	4.75

$$\gamma_1(100) + \gamma_2(100)^2 + \gamma_3(100)^3 = 0.5$$
$$\gamma_1(200) + \gamma_2(200)^2 + \gamma_3(200)^3 = 1.58$$
$$\gamma_1(400) + \gamma_2(400)^2 + \gamma_3(400)^3 = 3.75$$

Solving yields:

$$\gamma_1 = 0.658(10^{-3})/(°C)$$
$$\gamma_2 = 50.6(10^{-6})/(°C)^2$$
$$\gamma_3 = -72.0(10^{-9})/(°C)^3$$

Exercise 11.11

From Eq. (11.7):

$$\frac{\Delta R}{R_0} = 1 + \gamma_1(T - T_0) + \gamma_2(T - T_0)^2$$

$T - T_0$	100	400
$\Delta R/R_0$	1.33	2.42

$$\gamma_1(100) + \gamma_2(100)^2 = 0.33$$
$$\gamma_1(400) + \gamma_2(400)^2 = 1.42$$

Solving yields:

$$\gamma_1 = 3.217(10^{-3})/(°C)$$
$$\gamma_2 = 8.33(10^{-6})/(°C)^2$$

Exercise 11.12

From Eq. (11.5) if $\gamma_2 = \gamma_3 = 0$:

$$R = R_0(1 + \gamma_1 T)$$
$$= 100(1 + 0.003902\ T)$$

Case	Temp	R (Ω)
(a)	-240	6.35
(b)	-120	53.18
(c)	90	135.12
(d)	260	201.45
(e)	600	334.12
(f)	900	451.18

Exercise 11.13

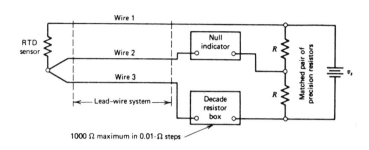

In the sensor circuit loop: $\quad R_S = R_{RTD} + R_{L1} + R_{L2}$

In the decode resistor loop: $\quad R_D = R_{DRB} + R_{L3} + R_{L2}$

But: $\quad R_{L1} = R_{L2} = R_{L3}$

When $R_{DRB} = R_{RTD}$: $\quad R_S = R_D$

No current flows through the null indicator regardless of the value of R_L.

Exercise 11.14

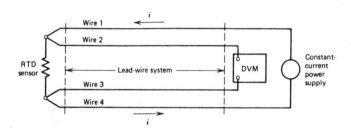

$$v_a - v_b = R_S(i - i_M) = i_M(2R_L + R_M)$$

$$= R_S i - R_S i_M = 2R_L i_M + R_M i_M$$

$$v_M = i_M R_M = R_S i - i_M(R_S + 2R_L)$$

But: $\quad i_M \approx \dfrac{i R_S}{R_M} \quad$ since $R_L \ll R_M$

Therefore: $\quad v_M = R_S i \left(1 - \dfrac{R_S}{R_M} - \dfrac{2R_L}{R_M}\right)$

Since R_m is of the order of 10^6 Ω or 10^7 Ω, load error (R_S/R_M) and lead wire error $(2R_L/R_M)$ are negligible.

Exercise 11.15

$$\frac{T}{T_m} = 1 - e^{-t/\beta}$$

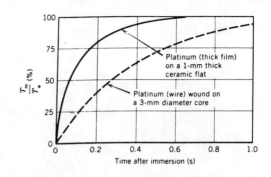

Type	T/T_m	t(s)	$e^{-t/\beta}$	t/β	$\beta = \frac{t}{1.609}$
thick film	0.8	0.20	0.2	1.609	0.124 s
wire	0.8	0.63	0.2	1.609	0.391 s

Exercise 11.16

$v = iR = 5(10^{-3})R$

Using the results from Exercise 11.12:

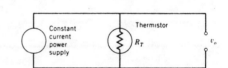

Case	Temp	R (Ω)	v (mV)
(a)	-240	6.35	32
(b)	-120	53.18	266
(c)	90	135.12	676
(d)	260	201.45	1007
(e)	600	334.12	1671
(f)	900	451.18	2256

Exercise 11.17

$v_B = 5$ V $\qquad F_{sh} = 0.35$ °C/mW

$i_B = v_B/R_e = v_B/R_T = 5/100 = 0.050$ A $= 50$ mA

$i_T = i_B/2 = 0.050/2 = 0.025$ A $= 25$ mA

$P_T = i_T^2 R_T = (0.025)^2(100) = 0.0625$ W $= 62.5$ mW

Error $= F_{sh} P_T = 0.35(62.5) = 21.9$°C

Exercise 11.18

$$\frac{\Delta R}{R_0} = 0.00392\left[T - 1.49\left(\frac{T}{100} - 1\right)\left(\frac{T}{100}\right) - \beta\left(\frac{T}{100} - 1\right)\left(\frac{T^3}{100}\right)\right]$$

$$\beta = 0 \text{ if } T > 0$$

$$\beta = 0.11 \text{ if } T < 0$$

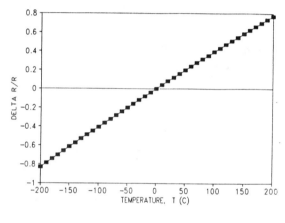

(a) $-200°C$ to $200°C$

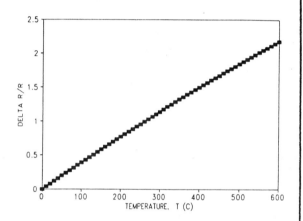

(b) $0°C$ to $600°C$

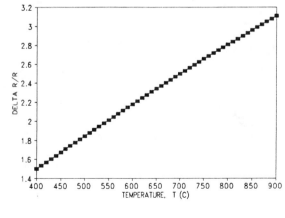

(c) $400°C$ to $900°C$

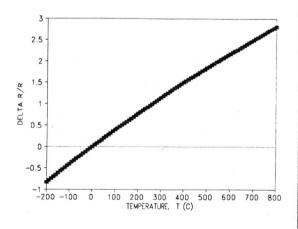

(d) $-200°C$ to $800°C$

Exercise 11.19

The four common errors are:

1. Lead wire effects: Minimize length, reduce R_L to 1% of R_T, use a 3- or a 4-lead wire system.

2. Stability: Do not employ sensors above their rated limit. If the limit is exceeded repeat measurements until stable readings are achieved.

3. Self heating: Adjust excitation voltage to limit power dissipation to 2 mW.

4. Sensitivity to strain: Strain effects are usually small but they can occur in bonded sensors. If possible, mount sensors in stress free regions or orient the grid in the minimum strain direction.

Exercise 11.20

For a thermistor:
$$\ln R/R_0 = \beta(1/\theta - 1/\theta_0)$$

Therefore:
$$R = R_0\, e^{\beta(1/\theta - 1/\theta_0)} = R_0\, e^{-\beta(\theta - \theta_0)/\theta\theta_0}$$

Since the range of a thermistor is relatively small, $\theta \approx \theta_0$.

Thus:
$$\frac{R}{R_0} \cong e^{-\beta\Delta\theta/\theta^2}$$

Differentiate with respect to $\Delta\theta$ and note that $\Delta\theta = \Delta T$:

$$S = \frac{d(R/R_0)}{d(\Delta\theta)} = -\frac{\beta}{\theta^2}\, e^{-\beta\Delta\theta/\theta^2}$$

For small $\Delta\theta$:
$$e^{-\beta\Delta\theta/\theta^2} \approx 1$$

Therefore:
$$S = -\beta/\theta^2$$

Exercise 11.21

From Eq. (11.12):
$$R = R_0 e^{\beta(1/\theta - 1/\theta_0)}$$

$$= 3000\, e^{4350(1/\theta - 1/298)}$$

Case	Temp (°C)	Temp (°K)	R (kΩ)
(a)	-80	193	8435
(b)	-40	233	176.06
(c)	0	273	11.42
(d)	50	323	0.969
(e)	75	348	0.368
(f)	150	423	0.040

Exercise 11.22

Thermistors are "resistance" type temperature sensors. Their advantages are low cost, small size, large output, and stability with time if they are encapsulated. The nonlinear response is accommodated by using signal conditioning circuits that incorporate a calibration table or the Steinhart-Hart relation to convert $\Delta R/R$ to T. The most significant disadvantage is the relatively low upper temperature limit of about 315°C. Stability is impaired if this temperature is exceeded.

RTDs are also "resistance" type temperature sensors. They are more costly, larger in size, and have a lower output than thermistors. Both sensors exhibit nonlinear output but the nonlinearities of the RTD are much smaller. The most significant advantage of the RTD is its capability for measuring very high temperature. With platinum sensors, temperatures as high as 1800°C can be measured.

Exercise 11.23

(a) 400°C   Use an RTD because of possible stability problems with a thermistor above 315°C.

(b) 90°C   Use a thermistor because of its higher output and lower cost.

(c) -200°C   Use an RTD because the resistance of the thermistor would be very high (See Exercise 11.21) and circuit noise would be a problem.

ENGINEERING MEASUREMENTS by J. W. DALLY, W. F. RILEY, AND K. G. McCONNELL

Exercise 11.24

For the constant-current potentiometer circuit:

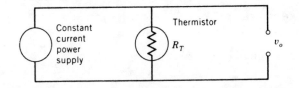

$$v_o = iR_T$$

From Eq. (11.12): $R_T = R_0 e^{\beta(1/\theta - 1/\theta_0)} = 3000\, e^{4350(1/\theta - 1/298)}$

For $i = 10$ mA: $v_o = iR_T = 0.010(3000) e^{4350(1/\theta - 1/298)}$

The output voltage v_o for $-50 \le \theta \le 300°C$ is shown below.

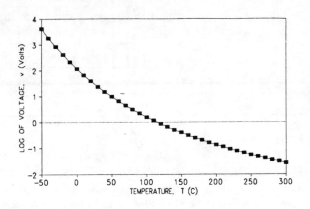

Exercise 11.25

From Eq. (11.14): $\dfrac{1}{\theta} = A + B \ln R_T + C (\ln R_T)^3$

(a) From Fig. 11.9 for $\beta = 3270$:

T	θ	$1/\theta$	R/R_0	R_T	$\ln R_T$	$(\ln R_T)^3$
25	298	0.003356	1	100	4.605	97.66
100	373	0.002681	0.12	12	2.485	15.34
175	448	0.002232	0.026	2.6	0.956	0.872

Thus:
$$A + 4.605\,B + 97.66\,C = 0.003356$$
$$A + 2.485\,B + 15.34\,C = 0.002681$$
$$A + 0.956\,B + 0.872\,C = 0.002232$$

Solving yields:

$A = 1.989(10^{-3})$ $B = 2.694(10^{-4})$ $C = 1.294(10^{-6})$

Exercise 11.25 (Continued)

(b) From Fig. 11.9 for $\beta = 4240$:

T	θ	$1/\theta$	R/R_o	R_T	$\ln R_T$	$(\ln R_T)^3$
25	298	0.003356	1	100	4.605	97.66
100	373	0.002681	0.06	6	1.792	5.75
175	448	0.002232	0.008	0.8	-0.223	-0.01

Thus:
$$A + 4.605\,B + 97.66\,C = 0.003356$$
$$A + 1.792\,B + 5.75\,C = 0.002681$$
$$A - 0.223\,B - 0.01\,C = 0.002232$$

Solving yields:

$A = 2.282(10^{-3})$ $B = 2.214(10^{-4})$ $C = 0.566(10^{-6})$

(c) From Fig. 11.9 for $\beta = 4710$:

T	θ	$1/\theta$	R/R_o	R_T	$\ln R_T$	$(\ln R_T)^3$
25	298	0.003356	1	100	4.605	97.66
100	373	0.002681	0.045	4.5	1.504	3.40
175	448	0.002232	0.0048	0.48	-0.734	-0.40

Thus:
$$A + 4.605\,B + 97.66\,C = 0.003356$$
$$A + 1.504\,B + 3.40\,C = 0.002681$$
$$A - 0.734\,B - 0.40\,C = 0.002232$$

Solving yields:

$A = 2.379(10^{-3})$ $B = 1.997(10^{-4})$ $C = 0.592(10^{-6})$

Exercise 11.26

$R_T = 5000\ \Omega$ $\mathcal{E} = 0.5\ °C$ $P = \mathcal{E}/F_{sh}$

$i = \sqrt{P/R_T} = \sqrt{\mathcal{E}/F_{sh} R_T}$

(a) For $F_{sh} = 0.5\ °C/mW$: $i = \sqrt{\mathcal{E}/F_{sh} R_T} = \sqrt{0.5/(500)(5000)} = 0.447$ mA

(b) For $F_{sh} = 1.0\ °C/mW$: $i = \sqrt{\mathcal{E}/F_{sh} R_T} = \sqrt{0.5/(1000)(5000)} = 0.316$ mA

(c) For $F_{sh} = 2.0\ °C/mW$: $i = \sqrt{\mathcal{E}/F_{sh} R_T} = \sqrt{0.5/(2000)(5000)} = 0.224$ mA

Exercise 11.27

Liquid-in-glass thermometers:

    Advantages: Low cost, easy to read, no instrumentation required.

    Disadvantages: Easily broken, time to reach equilibrium excessive, immersion required, can not be used for closed-loop control.

Bimetallic Thermometers:

    Advantages: Inexpensive, good for switch type control, available in a wide variety of shapes.

    Disadvantages: Time to reach equilibrium excessive, low accuracy, low sensitivity, poor dynamic response.

Pressure Thermometers:

    Advantages: Simple, low cost, permits remote sensing.

    Disadvantages: Limited accuracy, poor dynamic response.

Exercise 11.28

From Eq. (11.18) with the coefficients listed in Table A.6:

$$f(v) = T_1 - T_2 = a_0 + a_1 v_0 + a_2 v_0^2 + \cdots + a_n v_0^n$$

(a)   Chromel-Alumel(Type K):

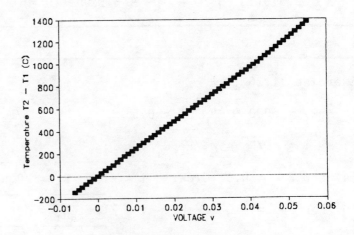

Exercise 11.28 (Continued)

(b) Chromel-constantan (Type E):

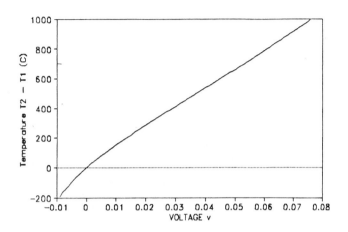

(c) Copper-constantan (Type T):

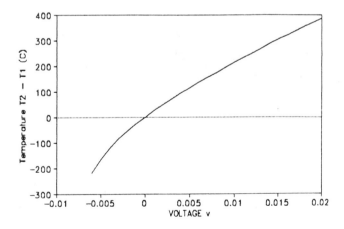

Exercise 11.29

(a) Chromel-Alumel (Type K)   Valid range $0 \leq T \leq 1370$ °C

Voltage	Table (°C)	Calculation (°C)
-6 mV	-207.5	-144.3 ← out of range
0 mV	0	0.2
40 mV	967.5	967.6
54 mV	1346.4	1346.3

(b) Chromel-Constantan (Type E):   Valid range $-100 \leq T \leq 1000$ °C

Voltage	Table (°C)	Calculation (°C)
-5 mV	-94.8	-94.8
25 mV	350.5	350.5
50 mV	661.1	661.1
75 mV	981.9	982.0

(b) Copper-Constantan (Type T):   Valid range $-160 \leq T \leq 400$ °C

Voltage	Table (°C)	Calculation (°C)
-4 mV	-123.0	-123.3
5 mV	115.3	115.3
12 mV	249.7	249.6
20 mV	385.9	385.7

Exercise 11.30

(a) From Table A.4:

$v_o = 14.570$ mV

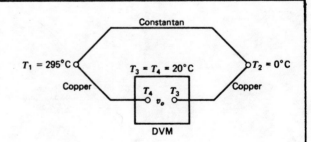

(b) From Table A.4:

$v_o = 2.078$ mV corresponds to $T_1 = 50.1$ °C.

(c) T_2 will influence the reading since it is the reference junction temperature and the tables are based on $T_2 = 0$ °C. T_2 or T_4 will not influence the reading because they are equal and the thermoelectric voltages produced will cancel.

Exercise 11.31

From Table A.4:

$v_o = 17.816$ mV

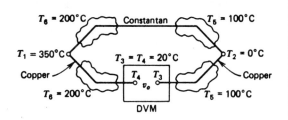

Exercise 11.32

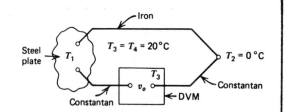

(a) From Table A.2:

$v_o = 25.412$ mV

(b) From Table A.2:

$v_o = 20.470$ mV corresponds to $T_1 = 496$ °C.

(c) No effect since $T_3 = T_4 = 20$ °C.

Exercise 11.33

(a) From Table A.5:

$T_1 = 260 + \dfrac{14.123 - 14.108}{14.163 - 14.108}$

$= 260.3$ °C

(b) Not if the temperature of the steel plate is uniform.

(c) No practical limit if the material properties remain the same and the temperature is uniform. The resistance of the plate may affect the instrument system. Usually junctions are maintained in close proximity (about 10 mm or less).

Exercise 11.34

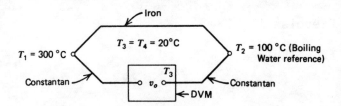

(a) From Table A.5:

$v_o = v_{300} - v_{100}$

$= 16.325 - 5.268 = 11.057$ mV

(b) $v_T = v_o + v_{100} = 21.333 + 6.268 = 26.801$ mV

$v_o = 26.801$ mV corresponds to $T_1 = 489.5$ °C.

Exercise 11.35

$R_L = 40(0.357) = 14.28\ \Omega$
which is negligible with
respect to $R_m = 10(10^6)\ \Omega$.

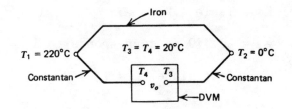

From Table A.2:

$v_T = 11.887$ mV

(a) $i = \dfrac{11.887(10^{-3})}{10(10^6)} = 1.1887(10^{-9})$ A $= 1.1887$ nA

$v_d = iR_L = 1.1887(10^{-9})(14.28) = 16.97(10^{-9})$ V $= 16.97$ nV

(b) $v_o = v_T - v_d = 11.887(10^{-3}) - 16.97(10^{-9})$

$= 11.887(10^{-3})$ V $= 11.887$ mV

(c) $v_o = 11.887$ mV corresponds to $T_1 = 220$°C

Exercise 11.36

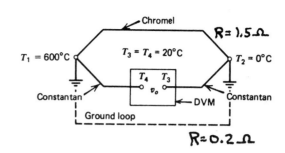

From Table A.2:

$v_T = 45.085$ mV

$R_e = \dfrac{(0.2)(1.5)}{0.2 + 1.5} = 0.176$ Ω

$i = v_T/R_e = \dfrac{45.085(10^{-3})}{0.176} = 256(10^{-3})$ A $= 256$ mA

$i_m = v_T/R_m = \dfrac{45.085(10^{-3})}{10(10^6)} = 4.51(10^{-9})$ A $= 4.51$ nA

$v_m = v_T - i_m R_T = 45.085(10^{-3}) - 4.51(10^{-9})(1.5)$

$\qquad = 45.085(10^{-3})$ V $= 45.085$ mV

The ground does not markedly affect the meter reading; however, the ground may contribute to noise pick-up. Also, the increased current produces an error due to the Peltier effect (see Eq. 11.16).

Exercise 11.37

For the circuit shown in Fig. E11.37:

$v = e_1 + e_5 + e_2 + e_4 + e_3 + e_5$

Since $T_3 = T_4$: $\quad e_3 = -e_4$

$v = e_1 + e_5 + e_2 + e_5$

$\quad = -16.35 + 2.058 + 0 + 2.058 = -12.234$ mV

Table A.5 indicates that no such reading is possible for an iron-constantan thermocouple. This fact is an indication that the thermocouple is not wired properly. If the 12.234 mV reading is taken as positive:

$v_o = 12.234$ mV corresponds to $T_1 = 226\ °$C which is in error by $74°$.

Exercise 11.38

While many different material combinations can be employed in forming thermocouples, seven different material pairs have been "standardized".

E Chromel-Constantan: A range from $-200°C$ to $1000°C$ with the highest sensitivity.

J Iron-Constantan: A range from $-185°C$ to $870°C$ with a very high sensitivity.

K Chromel-Alumel: A range from $-185°C$ to $1260°C$ with lower sensitivity than type E.

N Nicrosil-Nisil: A range from $-270°C$ to $1300°C$. Stability with time is enhanced.

R Pt-13% Rh-Pt: A range from $0°C$ to $1590°C$. Platinum and Platinum-Rhodium alloys are expensive.

S Pt-10% Rh-Pt: A range from $0°C$ to $1535°C$. Platinum and Platinum-Rhodium alloys are expensive.

T Copper-Constantan: Range is limited ($-185°C$ to $400°C$); however, a very popular combination for lower temperature laboratory measurements. Low cost, easy to form.

Exercise 11.39

From Fig. 11.17:

For $T = 777\ °C$, $t = 500$ h and # 14 wire:

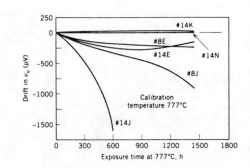

(a) Type J: Drift = $-1100\ \mu V = -1.1$ mV

$777°C$ is above the upper limit for J type thermocouples.
1.1 mV is equivalent to about an $18\ °C$ error in this range.

(b) Type E: Drift = $-250\ \mu V = -0.25$ mV

$v_o = 59.214 \quad T_1 = 777\ °C$
$v_o = 58.964 \quad T_1 = 773.8\ °C \quad\quad \text{error} = 3.2\ °C$

(c) Type K: Slight positive drift; negligible error.

(d) Type N: No drift; no error

Exercise 11.40

For Type N, # 8 wire,

$T = 1202\ °C$, $t = 1100$ h:

From Fig. 11.18:

Drift = 80 μV

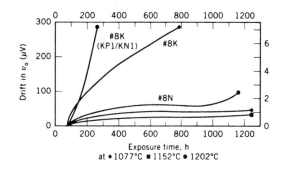

From Eq. (11.20):

$(v_o)_{A/B} = (v_o)_{A/C} - (v_o)_{B/C}$

$S_{Nicrosil/Nisil} = 15.4 - (-10.7) = 26.1\ \mu V/°C$

80 μV drift is equivalent to an error of $80/26.1 = 3.07\ °C$

Exercise 11.41

The simplest technique for controlling reference junction temperature utilizes an ice and water bath. The reference junction is immersed in a mixture of ice and water in a thermos bottle. Such an ice bath can maintain the water temperature to within $0.1\ °C$ ($0.2\ °F$) of the freezing point of water.

A very-high-quality reference temperature source employs thermoelectric refrigeration. Thermocouple wells in this unit contain distilled, water that is maintained at precisely $0\ °C$ ($32°\ F$). The outer walls of the wells are cooled by the thermoelectric refrigeration elements until freezing of the water in the wells begins. The increase in volume of the water as it begins to freeze expands a bellows that contacts a microswitch and deactivates the refrigeration elements. The cyclic freezing and thawing of the ice accurately maintains the temperature of the wells at $0\ °C$ ($32\ °F$). This automatic precise control of temperature can be maintained over extended periods of time.

The electrical-bridge method, incorporates a Wheatstone bridge, with a resistance temperature detector (RTD) as the active element, into the thermocouple circuit. The RTD and the reference junctions of the thermocouple are mounted on a reference block that is free to follow the ambient temperature. As the ambient temperature of the reference block varies, the RTD changes resistance. The bridge is designed to produce an output voltage that is equal but opposite to the voltage developed in the thermocouple circuit as a result of changes in the reference temperature. This method is widely used with potentiometric recording devices that are used to display one or more temperatures over long periods of time when it is obviously not practical to maintain the simple ice bath.

ENGINEERING MEASUREMENTS by J. W. DALLY, W. F. RILEY, AND K. G. McCONNELL

Exercise 11.42

Common instruments used to measure the output voltage from a thermocouple are the DVM, a strip chart recorder, and an oscilloscope. All of these instruments have a high input impedance that limits current flow in the thermocouple circuit and minimizes Peltier and Thompson effects.

An adaptation of the DVM contains a microprocessor to linearize the output so that the display reads directly in units of temperature instead of voltage. Also, cold junction compensation is employed to eliminate the need for a reference junction.

The strip chart recorder is employed for quasi-static measurements of temperature over long periods of time. These instruments also employ cold junction compensation methods. The scales are calibrated to read directly in units of temperature for a specified type of thermocouple. The chart provides a permanent record of temperature with time.

The oscilloscope is used only when the temperature fluctuations are at high frequency (several Hz or more). The impedance and sensitivity of the oscilloscope are well matched to the output of a thermocouple circuit.

Exercise 11.43

Cold junction compensation is accomplished by inserting an isothermal block within the instrument that serves as the mounting for the reference junction. The temperature of the block (which varies with the ambient temperature) is measured with a thermistor. The output from the thermistor is converted to a voltage signal, digitized, and subtracted from the voltage from the thermocouple circuit by the microprocessor in the instrument. Linearization in a temperature indicating DVM is accomplished by using the same microprocessor. The output voltage from the thermocouple circuit is digitized and used as input in Eq. (11.18). The temperature is computed by the microprocessor and then displayed.

Exercise 11.44

Two techniques can be employed to minimize noise. Shielded lead wires should be used so that noise due to capacitive-coupled fields is generated in the shields but not in the lead wires. The shield be connected to the guard terminal on the thermocouple DVM.

A second technique is to filter the noise. In most instances, the noise frequency is 60 Hz (50 Hz in some countries) due to common power lines. If a low pass filter is incorporated at the input terminals with a cut off frequency of 10 Hz, then the noise signals will be filtered out. If the temperature fluctuations occur slowly with frequencies of only a few Hz, the input signal is not attenuated.

Exercise 11.45

This brief should stress the primary advantage of the IC sensor:- that is, it acts as a constant current source where i is a linear function of temperature T. As a result, the IC sensor does not have problems of low signal output and there is no need for precision amplifiers, linearization circuits, and cold junction compensation. Most of the problems associated with the use of thermocouples, RTDs, and thermistors do not exist.

The most severe disadvantage of the IC sensor is its limited range of temperatures ($55°C$ to $150°C$).

Exercise 11.46

From Eq. (11.23): $\quad \dfrac{dT}{dt} + \dfrac{1}{\beta} T = \dfrac{1}{\beta} T_m \qquad$ where $\dfrac{1}{\beta} = \dfrac{hA}{mc}$

The homogeneous solution is: $\quad T = C_1 e^{-t/\beta}$

The particular solution is: $\quad T = T_m$

The general solution is: $\quad T = C_1 e^{-t/\beta} + T_m$

Since $T = 0$ at $t = 0$: $\quad C_1 = -T_m$

Thus: $\quad T = T_m \left[1 - e^{-t/\beta} \right]$

Exercise 11.47

From Eq. (11.23): $\quad \dfrac{dT}{dt} + \dfrac{1}{\beta} T = \dfrac{1}{\beta} T_m \qquad$ where $\dfrac{1}{\beta} = \dfrac{hA}{mc}$

For the special case where $T_m = bt$: $\quad \dfrac{dT}{dt} + \dfrac{1}{\beta} T = \dfrac{b}{\beta} t$

The homogeneous solution is: $\quad T = C_1 e^{-t/\beta}$

The particular solution is: $\quad T = b(t - \beta)$

The general solution is: $\quad T = C_1 e^{-t/\beta} + b(t - \beta)$

Since $T = 0$ at $t = 0$: $\quad C_1 = b\beta$

Thus: $\quad T = bt - b\beta \left[1 - e^{-t/\beta} \right]$

ENGINEERING MEASUREMENTS by J. W. DALLY, W. F. RILEY, AND K. G. McCONNELL

Exercise 11.48

For the ramp-hold input:

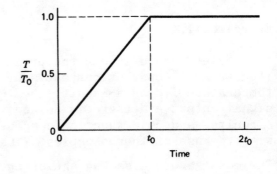

$\dfrac{T}{T_0} = bt = \dfrac{t}{t_0}$ For $0 < t \le t_0$

$T = T_0$ For $t > t_0$

From Eq. (11.29):

For $0 < t < t_0$: $\qquad \dfrac{T}{T_0} = b\beta e^{-t/\beta} + b(t - \beta)$

To hold at time t_0, add a negative ramp. Thus,

$$\dfrac{T}{T_0} = \left[b\beta e^{-t/\beta} + b(t - \beta)\right] - \left[b\beta e^{-(t-t_0)/\beta} + b(t - t_0 - \beta)\right]$$

For $0 < t < t_0$: $\qquad \dfrac{T}{T_0} = \dfrac{\beta}{t_0} e^{-t/\beta} + \left(\dfrac{t}{t_0} - \dfrac{\beta}{t_0}\right)$

For $t > t_0$: $\qquad \dfrac{T}{T_0} = 1 + \dfrac{\beta}{t_0} e^{-t/\beta} \left(1 - e^{t_0/\beta}\right)$

Exercise 11.49

The most difficult part of this calibration is finding a temperature source or bath that can be used to produce a linear variation of temperature with time. Once this problem is solved by using a calibrated cam-controlled electric oven or its equivalent, a temperature-time curve for the thermocouple can be obtained and used to establish the lag time as illustrated in Fig. 11.33 of the text.

Exercise 11.50

$T_f = 0.9 T_s \qquad \dfrac{h_s r_1}{k_s} = 1 \qquad \dfrac{\sqrt{(kA)_e/R}\ \tanh(L/\sqrt{(kA)_e/R})}{\pi r_1 k_s} = 8$

From Fig. 11.35: $\quad \dfrac{T_s - T}{T_s - T_f} = \dfrac{T_s - T}{T_s(1 - 0.9)} = 10\left[1 - \dfrac{T}{T_s}\right] = 0.72$

$\dfrac{T}{T_s} = 0.928 \qquad \mathscr{E} = \left[\dfrac{1 - 0.928}{1}\right](100) = 7.20\%$

Exercise 11.51

From Fig. 11.35:

(a) For $\dfrac{h_s r_1}{k_s} = 2$: $\quad \dfrac{T}{T_s} = 0.941 \quad \mathscr{E} = \left[\dfrac{1 - 0.941}{1}\right](100) = 5.90\%$

(b) For $\dfrac{h_s r_1}{k_s} = 4$: $\quad \dfrac{T}{T_s} = 0.962 \quad \mathscr{E} = \left[\dfrac{1 - 0.962}{1}\right](100) = 3.80\%$

(c) For $\dfrac{h_s r_1}{k_s} = 10$: $\quad \dfrac{T}{T_s} = 1.019 \quad \mathscr{E} = \left[\dfrac{1 - 1.019}{1}\right](100) = -1.90\%$

Exercise 11.52

From Eq. (11.31):

$T_v = \dfrac{V^2}{2 J g c_p}$

$c_p = 6000 \text{ ft·lb/slug·°R}$

$g = 32.2 \text{ ft/s}^2$

$J = 778 \text{ ft·lb/Btu}$

$T_v = \dfrac{V^2}{12{,}000} \quad (V \text{ in mph})$

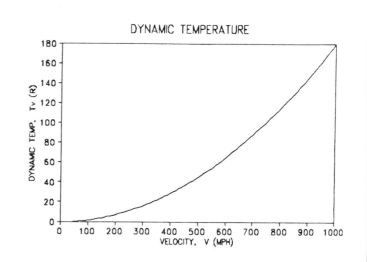

Exercise 11.53

This brief should describe the comparison method since it is relatively easy to perform if the laboratory has a standard sensor. The standard sensor and the thermocouples to be calibrated are mounted in close proximity on an isothermal block (an aluminum or copper plate 1 × 3 × 3 in.). The block is placed on a hotplate and its temperature is increased slowly (to maintain equilibrium). Readings at 5°F or 10°F increments are made from both sensors over the temperature range from 125°F to 600°F. Comparisons between the thermocouple output and the standard indicate the error and provide data to correct the errors.

Exercise 11.54

From Eq. (11.34):

$$E_\lambda = \frac{2\pi c^2 h}{\lambda^5 (e^{hc/k\lambda\theta} - 1)} = \frac{C_1}{\lambda^5 (e^{C_2/\lambda\theta} - 1)} = \frac{3.75(10^{-16})}{\lambda^5 (e^{0.0144/\lambda\theta} - 1)}$$

Log E_λ (W/m^2)

λ μm	100°C 373°K	200°C 473°K	500°C 773°K	1000°C 1273°K	2000°C 2273°K
0.1	-148.09	-112.64	-61.33	-29.55	-7.94
0.2	-65.76	-48.04	-22.38	-6.49	4.31
0.5	-17.45	-10.36	-0.10	6.25	10.58
1.0	-2.19	1.35	6.48	9.66	11.82
2.0	4.69	6.46	9.02	10.61	11.71
5.0	7.73	8.44	9.47	10.14	10.67
10.0	7.91	8.27	8.84	9.25	9.63
20.0	7.30	7.51	7.88	8.19	8.50
50.0	6.01	6.16	6.42	6.67	6.95
100.0	4.90	5.02	5.26	5.50	5.76

A graph of the above data is shown below.

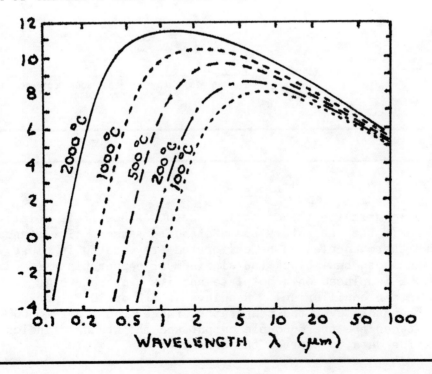

Exercise 11.55

From Eq. (11.35): $$\lambda_p = \frac{2898}{\theta} \, \mu m$$

$T(^\circ C)$	$\theta(^\circ K)$	Equation λ_p (μm)	Graph λ_p (μm)
100	373	7.77	8
200	473	6.13	6
500	773	3.75	4
1000	1273	2.28	2
2000	2273	1.27	1

Exercise 11.56

For: $T = 1000^\circ C = 1273 \text{ K}$

From Eq. (11.36):
$$E_t = 5.67(10^{-8}) \, \theta^4$$
$$= 5.67(10^{-8})(1273)^4$$
$$= 14.9(10^4) \text{ W/m}^2$$

Numerical integration of the area under the $1000^\circ C$ curve shown in Exercise 11.54 from 0.1 to 100 μm yields:

$$E_t = 14.3(10^4) \text{ W/m}^2$$

Exercise 11.57

From Eq. (11.38):
$$\theta = \frac{1}{\lambda_r (\ln \epsilon)/C_2 + 1/\theta_f}$$

$$C_2 = 1.44(10^{-2}) \text{ m} \cdot \text{K}$$

With $\lambda_r = 0.63(10^{-6})$ m:

θ °K	$\epsilon = 0.1$ θ_f °K	$\epsilon = 0.2$ θ_f °K	$\epsilon = 0.3$ θ_f °K	$\epsilon = 0.4$ θ_f °K	$\epsilon = 0.6$ θ_f °K	$\epsilon = 0.8$ θ_f °K
1000	1112	1076	1056	1042	1023	1010
1500	1767	1677	1629	1596	1552	1522
2000	2505	2328	2236	2174	2094	2040
2500	3342	3034	3879	2779	2648	2563
3000	4299	3803	3563	3410	3216	3091
3500	5406	4645	4291	4071	3797	3624
4000	6700	5568	5068	4764	4393	4163

A graph of the above data is shown below.

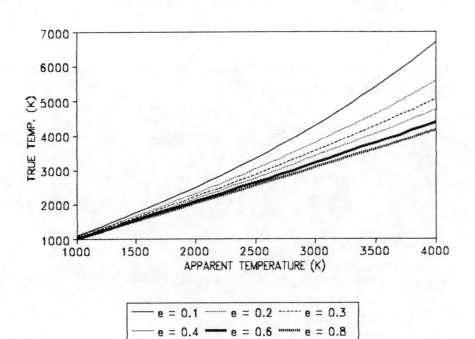

Exercise 11.58

From Eq. (11-37):
$$\frac{\varepsilon}{e^{(C_2/\lambda_r \theta)} - 1} = \frac{1}{e^{(C_2/\lambda_r \theta_f)} - 1}$$

If: $e^{(C_2/\lambda_r \theta)} \gg 1$ then $e^{C_2/\lambda_r \theta_f} \gg 1$

And
$$\varepsilon = \frac{e^{(C_2/\lambda_r \theta)}}{e^{(C_2/\lambda_r \theta_f)}}$$

$$\ln \varepsilon = \ln e^{(C_2/\lambda_r \theta)} - \ln e^{(C_2/\lambda_r \theta_f)}$$

$$= \frac{C_2}{\lambda_r \theta} - \frac{C_2}{\lambda_r \theta_f}$$

Thus:
$$\frac{1}{\theta} = \left(\ln \varepsilon + \frac{C_2}{\lambda_r \theta_f}\right) \frac{\lambda_r}{C_2}$$

$$\theta = \frac{1}{\lambda_r (\ln \varepsilon)/C_2 + 1/\theta_f}$$

$$\frac{d\theta}{\theta} = -\frac{(\lambda_r/C_2)(1/\varepsilon)\, d\varepsilon}{\lambda_r (\ln \varepsilon)/C_2 + 1/\theta_f} = -\frac{\lambda_r}{C_2} \theta \frac{d\varepsilon}{\varepsilon}$$

Exercise 11.59

From Eq. (11.39):

$$\frac{d\theta}{\theta} = -\frac{\lambda_r}{C_2} \theta \frac{d\varepsilon}{\varepsilon} = -\frac{0.63(10^{-6})}{1.44(10^{-2})} \theta \frac{d\varepsilon}{\varepsilon} = -4.375(10^{-5}) \theta \frac{d\varepsilon}{\varepsilon}$$

Error dT/T %

T °C	θ °K	$\frac{d\varepsilon}{\varepsilon} = 0.05$	$\frac{d\varepsilon}{\varepsilon} = 0.10$	$\frac{d\varepsilon}{\varepsilon} = 0.20$	$\frac{d\varepsilon}{\varepsilon} = 0.50$
1000	1273	0.28	0.56	1.11	2.78
1500	1773	0.39	0.78	1.55	3.88
2000	2273	0.50	0.99	1.99	4.97
2500	2273	0.61	1.21	2.43	6.07
3000	3273	0.72	1.43	2.86	7.16
3500	3773	0.83	1.65	3.30	8.25
4000	4273	0.93	1.87	3.74	9.35

Exercise 11.59 (Continued)

A graph of the above data is shown below.

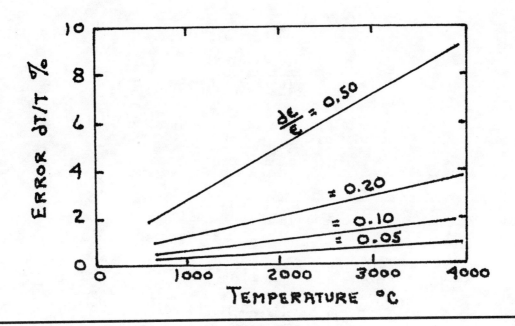

Exercise 11.60

The infrared pyrometer is an optical instrument that permits surface temperatures to be determined without contacting the heated body. The instrument incorporates a lens that collects the infrared radiation emitted from a focus spot on the body. This radiation is focused on a temperature sensor either a thermocouple or a thermistor. The equilibrium temperature of this sensor indicates the magnitude of the radiation from the surface and this magnitude is related to the surface temperature. The emissivity of the surface of the heated body affects the measurement. However, the instruments are equipped with emissivity compensation to correct for the reduction in radiant energy when $\varepsilon < 1$.

Exercise 11.61

From Eq. (11.42): $\qquad v_o = K\varepsilon T^3$

At $T = 400°C$: $\qquad v_o/v_0 = 10^4$

With $\varepsilon = 1$: $\qquad K = v_o/T^3 = 10^4 \, v_0/400^3 = 1.56(10^{-4})v_0$

Exercise 11.62

Optical pyrometers are thermal detectors which use radiation to heat a sensor. As such they are relatively slow because the sensor must be in thermal equilibrium.

A proton detector is a sensor that responds by generating a voltage that is proportional to the photon flux impinging on the photon sensor. These sensors respond several orders of magnitude faster than optical pyrometers and can be used for dynamic measurements at a single point or for field measurements of temperature. For field measurements, two oscillating lenses are used and a region of the body is scanned. The field is represented with many different temperature measurements to give a frame. A typical frame contains 28,000 measurements (280 lines of data with 100 elements per line).

Exercise 12.1

For steady laminar flow:

From Eq. (12.3): $\quad \dot{m} = \rho \int_A V \, dA = \rho V_{av} A$

From Eq. (12.2): $\quad V = V_0 \left(1 - \dfrac{r^2}{R^2}\right)$

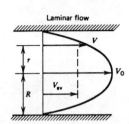

$$\dot{m} = \rho V_0 \int_0^{2\pi} \int_0^R \left(1 - \frac{r^2}{R^2}\right) r \, dr \, d\theta = \frac{\rho V_0}{R^2} \int_0^{2\pi} \int_0^R (R^2 r - r^3) \, dr \, d\theta$$

$$= \frac{2\pi \rho V_0}{R^2} \left[\frac{R^2 r^2}{2} - \frac{r^4}{4}\right]_0^R = \frac{\rho \pi R^2 V_0}{2}$$

Since the cross-sectional area of a circular pipe is πR^2,

$$\dot{m} = \frac{\rho A V_0}{2} = \rho A V_{av} \qquad \text{therefore} \qquad V_{av} = \frac{V_0}{2}$$

Exercise 12.2

For fully developed flow:

From Eq. (12.5): $\quad V = V_0 \left(1 - \dfrac{r}{R}\right)^{1/n}$

From Eq. (12.3): $\quad \dot{m} = \rho \int_A V \, dA = \rho V_{av} A$

$$\dot{m} = \rho V_0 \int_0^{2\pi} \int_0^R \left(1 - \frac{r}{R}\right)^{1/n} r \, dr \, d\theta = 2\pi \rho V_0 \int_0^R \left(1 - \frac{r}{R}\right)^{1/n} r \, dr$$

$$= 2\pi \rho V_0 R^2 \int_0^1 \left(1 - \frac{r}{R}\right)^{1/n} \left(\frac{r}{R}\right) d\left(\frac{r}{R}\right)$$

Let $x = \dfrac{r}{R}$:

$$= 2\pi \rho V_0 R^2 \int_0^1 (1 - x)^{1/n} x \, dx$$

$$= 2\pi \rho V_0 R^2 \frac{n^2}{(1 + 2n)(n + 1)}$$

Since the cross-sectional area of a circular pipe is πR^2,

$$\dot{m} = 2\rho A V_0 \frac{n^2}{(1 + 2n)(n + 1)} = \rho V_{avg} A$$

Therefore:

$$V_{avg} = \frac{2n^2}{(n + 1)(1 + 2n)} V_0$$

Exercise 12.3

From Table B-1:
$$\gamma = 62.37 \text{ lb/ft}^3$$
$$\rho = 1.938 \text{ slugs/ft}^3$$
$$\mu = 2.359(10^{-5}) \text{ lb·s/ft}^2$$

(a) $\dot{W} = \gamma A V = 62.37(\pi)(5/12)^2(20) = 680 \text{ lb/s}.$

(b) $\dot{m} = \rho A V = 1.938(\pi)(5/12)^2(20) = 21.1 \text{ slug/s}.$

(c) $E_T = \gamma Q V^2/2g = \dot{W} V^2/2g = \dfrac{680(20)^2}{2(32.2)} = 4224 \text{ ft·lb/s}$

(d) For a circular pipe:
$$R_e = \frac{\rho V_{av} D}{\mu} = \frac{1.938(20)(10/12)}{2.359(10^{-5})} = 1.369(10^6)$$

From Fig. 12.2: $\quad n = 9.8$

(e) From Eq. (12.6):
$$V_0 = \frac{(n+1)(2n+1)}{2n^2} V_{av} = \frac{10.8(20.6)}{2(9.8)^2}(20) = 23.2 \text{ ft/s}.$$

Exercise 12.4

From Eq. (12.3):
$$\dot{m} = \rho \int_A V \, dA = \rho V_{av} A$$

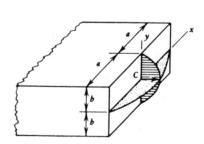

For the velocity profile:
$$V = V_C \left[1 - (x/a)^2\right]\left[1 - (y/b)^2\right]$$

$$\dot{m} = \rho V_C \int_{-a}^{+a} \int_{-b}^{+b} \left[1 - (x/a)^2\right]\left[1 - (y/b)^2\right] dx\, dy = \frac{16 \rho V_C a b}{9}$$

Since the cross-sectional area of the duct is $A = 2a(2b) = 4ab$:

$$\dot{m} = \frac{4 \rho V_C A}{9} = \rho V_{av} A$$

Therefore:
$$V_{av} = \frac{4}{9} V_C$$

Exercise 12.5

From Eq. (12.8):
$$\frac{p_o}{\gamma} + \frac{V_{oi}^2}{2g} = \frac{p_s}{\gamma} + \frac{V_s^2}{2g}$$

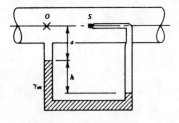

Since $V_s = 0$:
$$V_{oi}^2 = 2g\left(\frac{p_s - p_o}{\gamma}\right)$$

For the manometer:
$$p_o + \gamma a + \gamma_m h = p_s + \gamma(a + h)$$
$$p_s - p_o = (\gamma_m - \gamma)h$$

Therefore:
$$V_{oi}^2 = 2g\left(\frac{\gamma_m - \gamma}{\gamma}\right)h \qquad (a)$$

$$V_o = C_1 V_{oi} = C_1 \left[2g \frac{\gamma_m - \gamma}{\gamma} h\right]^{1/2} = K\sqrt{h}$$

From table B-2: $\gamma_{mercury} = 132.6 \text{ kN/m}^3$ $\gamma_{glycerin} = 12.33 \text{ kN/m}^3$

For: $C_1 = 0.98$ and h in centimeters

$$V_o = 0.98\sqrt{2(9.81)\frac{(132.6 - 12.33)}{12.33}\frac{h}{100}} = 1.356\sqrt{h}$$

Therefore: $K = 1.356$

For the inclined pipe:
$$\frac{p_o}{\gamma} + \frac{V_{oi}^2}{2g} + z_o = \frac{p_s}{\gamma} + \frac{V_s^2}{2g} + z_s$$

$$V_{oi}^2 = 2g\left(\frac{p_s - p_o}{\gamma} + z_s - z_o\right)$$

For the manometer:
$$p_o + \gamma a + \gamma_m h = p_s + \gamma(a + h + b)$$
$$p_s - p_o = (\gamma_m - \gamma)h - \gamma b$$

Therefore:
$$V_{oi}^2 = 2g\left(\frac{(\gamma_m - \gamma)h}{\gamma} - b + (z_s - z_o)\right)$$

Since $(z_s - z_o) = b$
$$V_{oi}^2 = 2g\frac{\gamma_m - \gamma}{\gamma}h \qquad (b)$$

Equations (a) and (b) are identical; therefore, no change results from the pipe inclination of 60°.

Exercise 12.6

From Eq. (12.9):

$$p_d = \frac{\gamma V_{ot}^2}{2g} \quad \text{(True dynamic pressure)}$$

$$p_d' = \frac{\gamma V_{om}^2}{2g} \quad \text{(Measured dynamic pressure)}$$

$$\mathcal{E} = \frac{p_d - p_d'}{p_d} = \frac{V_{ot}^2 - V_{om}^2}{V_{ot}^2}$$

Therefore:

$$V_{om}^2 = (1 - \mathcal{E}) V_{ot}^2$$

$$V_{om} = \sqrt{1 - \mathcal{E}} \; V_{ot}$$

From Fig. 12.7 at a 40° angle error: $\mathcal{E}_1$ (standard end) = 0.40

$\mathcal{E}_2$ (square end) = 0.11

$$V_{om1} = \sqrt{1 - 0.40} \; V_{ot} = 0.77 \; V_{ot}$$

$$V_{om2} = \sqrt{1 - 0.11} \; V_{ot} = 0.94 \; V_{ot}$$

The standard-end pitot tube would give a velocity about 23% low.

The square-end pitot tube would give a velocity about 6% low.

Exercise 12.7

From Table B-2:

For Glycerin: $\rho = 2.439$ slugs/ft^3 $\mu = 3120(10^{-5})$ lb·s/ft^2

For Water: $\rho = 1.936$ slugs/ft^3 $\mu = 2.10(10^{-5})$ lb·s/ft^2

From Fig. 12.8: $C_p > 1$ if Re < 100

For $\text{Re} = \frac{\rho V D}{\mu} = 100$: $V_{min} = \frac{100 \mu}{\rho D}$

For glycerin: $V_{min} = \frac{100 \mu}{\rho D} = \frac{100(3120)(10^{-5})}{2.439 (0.125/12)} = 122.8$ ft/s.

For water: $V_{min} = \frac{100 \mu}{\rho D} = \frac{100(2.10)(10^{-5})}{1.936 (0.125/12)} = 0.104$ ft/s.

Exercise 12.8

$$p_s - p_o = \gamma h = 62.4(35/12) = 182 \text{ lb/ft}^2 = 1.264 \text{ psi}$$

From Eq. (12.10):

$$V_o = \left[2g\left(\frac{p_s - p_o}{\gamma}\right)\right]^{1/2} = \left[2\left(\frac{p_s - p_o}{\rho}\right)\right]^{1/2}$$

For air at $T = 30°F$ and $p = 12.3$ psia:

$$\rho = \frac{p}{RT} = \frac{12.3(144)}{1715(460 + 30)} = 2.108(10^{-3}) \text{ slug/ft}^3$$

$$V_o = \left(\frac{2(182)}{2.108(10^{-3})}\right)^{1/2} = 416 \text{ ft/s} = 283 \text{ mi/h}$$

From Eq. (12.18):

$$V_o^2 = 2C_p T_s \left[1 - \left(\frac{p_o}{p_s}\right)^{(k-1)/k}\right]$$

$$p_s = 12.3 + 1.264 = 13.564 \text{ psia}$$

$$\frac{k-1}{k} = \frac{1.4 - 1}{1.4} = 0.2857$$

$$\left(\frac{p_o}{p_s}\right)^{(k-1)/k} = \left(\frac{12.3}{13.564}\right)^{0.2857} = 0.972$$

$$V_o = \sqrt{2(6000)(490)(1 - 0.972)} = 403 \text{ ft/s} = 274 \text{ mi/h}$$

$$\text{Difference} = \frac{283 - 274}{274} = 0.0328 = 3.28\%$$

Exercise 12.9

From Eq. (12.19):

$$(A + B\sqrt{\rho V})(\theta_a - \theta_f) = i_a^2 R_a$$

$$\sqrt{V} = \frac{i_a^2 R_a}{B\sqrt{\rho}(\theta_a - \theta_f)} - \frac{A}{B\sqrt{\rho}}$$

$$= \frac{R_a}{B\sqrt{\rho}(\theta_a - \theta_f)}\left[i_a^2 - \frac{A(\theta_a - \theta_f)}{R_a}\right]$$

$$V = \left(\frac{R_a}{B\sqrt{\rho}(\theta_a - \theta_f)}\right)^2 \left[i_a^2 - \frac{A(\theta_a - \theta_f)}{R_a}\right]^2$$

When $V = 0$:

$$i_a^2 = \frac{A(\theta_a - \theta_f)}{R_a} = i_o^2$$

Therefore:

$$V = \left(\frac{R_a}{B(\theta_a - \theta_f)\sqrt{\rho}}\right)^2 \left(i^2 - i_o^2\right)^2$$

$$= \left(\frac{R_a i_o^2}{B(\theta_a - \theta_f)\sqrt{\rho}}\right)^2 \left[\left(\frac{i}{i_o}\right)^2 - 1\right]^2$$

$$= C_o \left[\left(\frac{i}{i_o}\right)^2 - 1\right]^2$$

Where

$$i_o = \sqrt{A(\theta_a - \theta_f)/R_a}$$

$$C_o = \left(\frac{R_a i_o^2}{B(\theta_a - \theta_f)\sqrt{\rho}}\right)^2 = \left(\frac{A}{B\sqrt{\rho}}\right)^2 = \frac{A^2}{B^2 \rho}$$

Exercise 12.10

From Fig. 12.13:

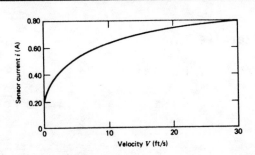

$i_0 = 0.20$ A   at   V = 0
$i = 0.80$ A   at   V = 30 ft/s

From Eq. (12.21):

$$V = C_0\left[\left(\frac{i}{i_0}\right)^2 - 1\right]^2 = C_0\left[\left(0.80/0.20\right)^2 - 1\right]^2 = 30$$

Thus: $C_0 = 30/225 = 0.1333$ ft/s

From the results of Problem 12.9 (with $\theta_a - \theta_f = 75°F$):

$$i_0^2 = \frac{A(\theta_a - \theta_f)}{R_a} \qquad \frac{A}{R_a} = \frac{i_0^2}{\theta_a - \theta_f} = \frac{(0.20)^2}{75} = 0.000533 \text{ A}^2/°F$$

For $\theta_a - \theta_f = (75 + 8)°F = 83°F$:

$$i_{0c}^2 = \sqrt{\frac{A}{R_a}(\theta_a - \theta_f)} = \sqrt{0.000533(83)} = 0.2104 \text{ A}^2$$

$$\frac{V_{True}}{V_{meas}} = \frac{\left[\left(\frac{i}{i_{0c}}\right)^2 - 1\right]^2}{\left[\left(\frac{i}{i_0}\right)^2 - 1\right]^2}$$

At i = 0.30 A:

$$V_{True} = \frac{\left[\left(\frac{i}{i_{0c}}\right)^2 - 1\right]^2}{\left[\left(\frac{i}{i_0}\right)^2 - 1\right]^2} V_{Meas} = \frac{\left[\left(\frac{0.30}{0.2104}\right)^2 - 1\right]^2}{\left[\left(\frac{0.30}{0.20}\right)^2 - 1\right]^2} V_{Meas} = 0.683 \, V_{Meas}$$

At i = 0.50 A:

$$V_{True} = \frac{\left[\left(\frac{i}{i_{0c}}\right)^2 - 1\right]^2}{\left[\left(\frac{i}{i_0}\right)^2 - 1\right]^2} V_{Meas} = \frac{\left[\left(\frac{0.50}{0.2104}\right)^2 - 1\right]^2}{\left[\left(\frac{0.50}{0.20}\right)^2 - 1\right]^2} V_{Meas} = 0.784 \, V_{Meas}$$

At i = 0.70 A:

$$V_{True} = \frac{\left[\left(\frac{i}{i_{0c}}\right)^2 - 1\right]^2}{\left[\left(\frac{i}{i_0}\right)^2 - 1\right]^2} V_{Meas} = \frac{\left[\left(\frac{0.70}{0.2104}\right)^2 - 1\right]^2}{\left[\left(\frac{0.70}{0.20}\right)^2 - 1\right]^2} V_{Meas} = 0.801 \, V_{Meas}$$

Exercise 12.11

From Table B-2:
$$\rho = 998 \text{ kg/m}^3$$
$$\mu = 1.00(10^{-3}) \text{ N·s/m}^2$$

From Fig. 12.15:
$$C_D = 1.05 \quad \text{For Re} > 3000$$
$$A = (\pi/4)(0.025)^2 = 0.491(10^{-3}) \text{ m}^2$$

From Eq. (12.22):
$$F_D = C_D \frac{\rho V^2 A}{2}$$
$$V = \left[\frac{2F_D}{C_D \rho A}\right]^{1/2} = S_v \sqrt{F_D}$$

Therefore:
$$S_v = \left[\frac{2}{C_D \rho A}\right]^{1/2} = \left[\frac{2}{1.05(998)(0.491)(10^{-3})}\right]^{1/2}$$
$$= 1.972 \text{ m/(s·}\sqrt{N}\text{)}$$
$$V > \frac{\text{Re } \mu}{\rho D} > \frac{3000(1.00)(10^{-3})}{998(0.025)} = 0.1202 \text{ m/s}$$

Exercise 12.12

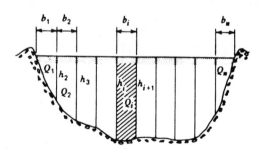

From Eq. (12.26):
$$V_i = \frac{1}{2}\left(V_{2i} + V_{8i}\right)$$
$$= \frac{2.45}{2(60)}\left[N_{0.2h} + N_{0.8h}\right]$$

From Eq. (12.27a):
$$Q_i = \frac{1}{4} b_i (h_i + h_{i+1})(V_i + V_{i+1})$$

Exercise 12.12 (Continued)

V_{i+1}	$h_i + h_{i+1}$	$V_i + V_{i+1}$	Q_i
1.456	3.0	1.456	13.10
1.891	6.5	3.347	65.27
2.083	7.7	3.974	91.80
2.254	7.9	4.337	102.79
2.421	8.8	4.675	123.42
2.323	9.7	4.744	138.05
2.164	8.4	4.487	113.07
2.136	7.8	4.300	100.62
1.985	7.2	4.121	89.01
1.772	6.3	3.757	71.01
1.450	5.1	3.237	49.53
0	2.0	1.450	8.70

From Eq. (12.27b): $\qquad Q = \sum Q_i = 966 \text{ ft}^3/\text{s}$

Exercise 12.13

From Table B-2: $\quad \rho = 1.936 \text{ slug/ft}^3 \qquad \mu = 2.10(10^{-5}) \text{ lb·s/ft}^2$

From Eq. (12.28): $\qquad S_N = \dfrac{f_s D}{V}$

For $300 < \text{Re} < 150{,}000$: $\qquad 0.20 < S_N < 0.21$

$$f_s = \frac{S_N}{D} V = \frac{0.21}{0.250/12} V = 10.08\, V \cong 10\, V$$

$$\text{Re} = \frac{\rho V D}{\mu} = \frac{1.936(0.250/12)}{2.10(10^{-5})} V = 1921\, V$$

For $V = 0.2$ ft/s: $\qquad \text{Re} = 384$

For $V = 10$ ft/s: $\qquad \text{Re} = 19{,}210$

The cylinder should be mounted such that $f_n > 3 f_s > 300$ Hz.

Exercise 12.14

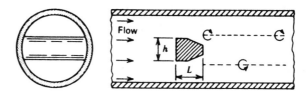

For $10{,}000 < Re < 1{,}000{,}000$:

$$S_N = 0.88 \pm 0.01$$
$$h = D/3 = 12/3 = 4.00 \text{ in.}$$
$$L = 1.3h = 1.3(4.00) = 5.20 \text{ in.}$$

From Eq. (12.28):

$$V_o = \frac{h}{S_N} f_s = \frac{4.00/12}{0.88} f_s = 0.379\, f_s = S_v f_s$$

Therefore:
$$S_v = 0.379 \text{ (ft/s)/Hz}$$

From Table B-2:
$$\rho = 1.936 \text{ slugs/ft}^3$$
$$\mu = 2.10(10^{-5}) \text{ lb·s/ft}^2$$

$$Re = \frac{\rho V h}{\mu} = \frac{1.936(4.0/12)}{2.10(10^{-5})} V = 30{,}730\, V$$

$$V_{min} > \frac{Re_{min}}{30730} = \frac{10{,}000}{30{,}730} > 0.325 \text{ ft/sec.}$$

$$V_{max} > \frac{Re_{max}}{30730} = \frac{1{,}000{,}000}{30{,}730} > 32.5 \text{ ft/sec.}$$

$$Q = AV = AS_v f_s = S_Q f_s$$

$$S_Q = AS_v = \left(\frac{\pi}{4}\right)(D)^2 S_v = \left(\frac{\pi}{4}\right)\left(\frac{12}{12}\right)^2 (0.379) = 0.298 \text{ (ft}^3\text{/s)/Hz}$$

Exercise 12.15

From Eq. (12.30):

$$\frac{p_1}{\gamma} + \frac{V_1^2}{2g} + z_1 = \frac{p_2}{\gamma} + \frac{V_2^2}{2g} + z_2$$

Also since $\rho_1 = \rho_2 = \rho$:

$$\dot{m} = \rho A_1 V_1 = \rho A_2 V_2$$

Therefore: $\qquad V_1 = \dfrac{A_2}{A_1} V_2$

$$\frac{V_2^2}{2g}\left[1 - \left(\frac{A_2}{A_1}\right)^2\right] = \frac{p_1 - p_2}{\gamma} - (z_2 - z_1) = \frac{p_1 - p_2}{\gamma} - a$$

For the manometer:

$$p_1 + \gamma(b) + \gamma(h) = p_2 + \gamma(a) + \gamma(b) + \gamma_m(h)$$

$$\frac{p_1 - p_2}{\gamma} = \frac{\gamma_m - \gamma}{\gamma} h + a$$

$$V_{2i}^2 = \frac{2g}{1 - (A_2/A_1)^2}\left(\frac{\gamma_m - \gamma}{\gamma}\right) h$$

$$Q_a = C_v Q_i = C_v V_{2i} A_2$$

$$Q_a = \frac{C_v A_2}{\left[1 - (A_2/A_1)^2\right]^{1/2}}\left[2g\left(\frac{\gamma_m - \gamma}{\gamma}\right) h\right]^{1/2}$$

(a) Q_a is independent of a & b.

(b) Place the manometer above the pipe.

(c) As a result of changes in the manometer equation,

$$\frac{\gamma_m - \gamma}{\gamma} \qquad \text{becomes} \qquad \frac{\gamma - \gamma_m}{\gamma}$$

Exercise 12.16

(a) From Eq. (12.32):

$$Q_i = \frac{A_2}{\left[1 - (A_2/A_1)^2\right]^{1/2}} \left[2g\left(\frac{p_1}{\gamma} + z_1 - \frac{p_2}{\gamma} - z_2\right)\right]^{1/2}$$

For the manometer:

$$p_1 + \gamma(b) + \gamma(h) = p_2 + \gamma(z_2 - z_1) + \gamma(b) + \gamma_m(h)$$

$$\frac{p_1 - p_2}{\gamma} = \left(\frac{\gamma_m - \gamma}{\gamma}\right) h + (z_2 - z_1)$$

Therefore:

$$Q_i = \frac{A_2}{\left[1 - (A_2/A_1)^2\right]^{1/2}} \left[2g\left(\frac{\gamma_m - \gamma}{\gamma}\right) h\right]^{1/2}$$

With h measured in inches of water:

$$Q_i = \frac{A_2}{\left[1 - (A_2/A_1)^2\right]^{1/2}} \left[2g\left(\frac{\gamma_m - \gamma}{\gamma}\right) \frac{h}{12}\right]^{1/2}$$

$$\frac{A_2}{A_1} = \frac{(\pi/4)d_2^2}{(\pi/4)d_1^2} = \frac{(1.5)^2}{(3.0)^2} = 0.25$$

$$A_2 = (\pi/4)(1.5)^2 = 1.767 \text{ in}^2 = 0.01227 \text{ ft}^2$$

From Table B-2: $\gamma_m = 62.32 \text{ lb/ft}^3$ $\gamma = 53.6 \text{ lb/ft}^3$

$$2g \frac{\gamma_m - \gamma}{\gamma} \frac{h}{12} = \frac{2(32.2)(62.32 - 53.6)h}{53.6(12)} = 0.873h$$

$$Q_i = S_{Qi} \sqrt{h}$$

$$S_{Qi} = \frac{Q_i}{\sqrt{h}} = \frac{A_2}{\left[1 - (A_2/A_1)^2\right]^{1/2}} \left[2g\left(\frac{\gamma_m - \gamma}{\gamma}\right) \frac{h}{12}\right]^{1/2}$$

$$= \frac{0.01227(\sqrt{0.873})}{\sqrt{1 - (0.25)^2}} = 0.01184 \text{ (ft}^3/\text{s)}/\sqrt{\text{in.}}$$

$$S_{Vi} = \frac{S_{Qi}}{A_2} = \frac{0.01184}{0.01227} = 0.965 \text{ (ft/s) } \sqrt{\text{in}}$$

Exercise 12.16 (Continued)

(b) From Table B-2:
$$\rho = 1.665 \text{ slug/ft}^3$$
$$\mu = 3.11(10^{-5}) \text{ lb·/ft}^2$$
$$V_{2i} = S_{Vi}\sqrt{h} = 0.965\sqrt{h}$$

Therefore:
$$Re = \frac{\rho V_2 d_2}{\mu} = \frac{1.665(0.965\sqrt{h})(1.5/12)}{3.11(10^{-5})} = 6458\sqrt{h}$$

For $h = 18$ in.:
$$Re = 6458\sqrt{18} = 28{,}000$$

From Fig. 12.27 with $D = 3.0$ in. and $Re = 28{,}000$: $\quad C_v \cong 0.96$

(c) $\quad Q_a = C_v Q_i = C_v S_{Qi}\sqrt{h}$
$$= 0.96(0.01184)(\sqrt{18}) = 0.0482 \text{ ft}^3/\text{s}$$

Exercise 12.17

$$A_2 = \frac{\pi}{4}d_2^2 = \frac{\pi}{4}\left(\frac{5}{12}\right)^2 = 0.1364 \text{ ft}^2$$

$$V_{2\,min} = \frac{Q_{min}}{A_2} = \frac{0.05}{0.1364} = 0.367 \text{ ft/s}$$

$$V_{2\,max} = \frac{Q_{max}}{A_2} = \frac{3.00}{0.1364} = 22.0 \text{ ft/s}$$

From Table B-1:

$\rho = 1.902 \text{ slug/ft}^3 \quad \mu = 0.905(10^{-5}) \text{ lb·s/ft}^2 \quad \gamma = 61.20 \text{ lb/ft}^3$

$$Re = \frac{\rho V_2 d_2}{\mu} = \frac{1.902(5/12)}{0.905(10^{-5})} V_2 = 87{,}570\, V_2$$

From Figure 12.27 for $d/D = 6/10 = 0.60$:

For $V_2 = 0.367$ ft/s:
$\quad Re = 87{,}570\, V_2 = 87{,}570(0.367) = 32{,}138 \qquad C_v = 0.958$

For $V_2 = 22.0$ ft/s:
$\quad Re = 87{,}570\, V_2 = 87{,}570(22.0) = 1{,}926{,}540 \qquad C_v = 0.992$

Exercise 12.17 (Continued)

From Eq. (12.33) with $z_1 = z_2$:

$$p_1 - p_2 = \frac{\gamma Q_a^2}{2g(C_v^2 A_2^2)}\left[1 - (A_2/A_1)^2\right]$$

$$\frac{A_2}{A_1} = \frac{(\pi/4)(6)^2}{(\pi/4)(10)^2} = 0.36 \qquad 1 - (A_2/A_1)^2 = 1 - (0.36)^2 = 0.8704$$

When $Q_a = 0.05$ ft^3/s:

$$p_1 - p_2 = \frac{61.2(0.05)^2(0.8704)}{2(32.2)(0.958)^2(0.1364)^2} = 0.1211 \text{ lb/ft}^2$$

When $Q_a = 3.00$ ft^3/s:

$$p_1 - p_2 = \frac{61.2(3.00)^2(0.8704)}{2(32.2)(0.992)^2(0.1364)^2} = 407 \text{ lb/ft}^2$$

From the manometer equation (see exercise 12.16):

$$\frac{p_1 - p_2}{\gamma} = \left[\frac{\gamma_m - \gamma}{\gamma}\right]\frac{h}{12} \qquad \text{or} \qquad h = \frac{12(p_1 - p_2)}{\gamma_m - \gamma}$$

With $\gamma_m = \gamma_{mercury} = 844$ lb/ft^3 (see table B-2):

$$h_H = \frac{12(p_1 - p_2)}{\gamma_m - \gamma} = \frac{12(407)}{844 - 61.2} = 6.24 \text{ in.}$$

$$h_L = \frac{12(p_1 - p_2)}{\gamma_m - \gamma} = \frac{12(0.1211)}{844 - 61.2} = 0.001856 \text{ in.}$$

With $\gamma_m = \gamma_{glycerin} = 78.5$ lb/ft^3 (see table B-2):

$$h_H = \frac{12(p_1 - p_2)}{\gamma_m - \gamma} = \frac{12(407)}{78.5 - 61.2} = 282 \text{ in.}$$

$$h_L = \frac{12(p_1 - p_2)}{\gamma_m - \gamma} = \frac{12(0.1211)}{78.5 - 61.2} = 0.084 \text{ in.}$$

A mercury manometer could be used for the high pressures but would not be satisfactory for the very low pressures. A glycerin manometer could be used for the low pressures but would require too large a height h for the high pressures.

Exercise 12.18

From Table B-2:

$$\gamma = 8.42 \text{ kN/m}^3 \qquad \rho = 858 \text{ kg/m}^3$$

$$\gamma_m = 132.6 \text{ kN/m}^3 \qquad \mu = 7.13(10^{-3}) \text{ N·s/m}^2$$

From Eq. (12.35):

$$Q_a = CA_0 \left[2g\left(\frac{p_1 - p_2}{\gamma} - (z_2 - z_1)\right) \right]^{1/2}$$

From the manometer equation (see Exercise 12.16):

$$\frac{\gamma_m - \gamma}{\gamma} h = \frac{p_1 - p_2}{\gamma} - (z_2 - z_1)$$

$$Q_a = CA_0 \left[2g\left(\frac{\gamma_m - \gamma}{\gamma}\right) h \right]^{1/2}$$

$$A_0 = (\pi/4)(d_0)^2 = (\pi/4)(0.030)^2 = 0.707(10^{-3}) \text{ m}^3$$

$$\frac{d}{D} = \frac{30}{100} = 0.30 \qquad C \cong 0.60 \text{ (From Fig. 12.29)}$$

$$Q_a = CA_0 \left[2g\left(\frac{\gamma_m - \gamma}{\gamma}\right) h \right]^{1/2}$$

$$= 0.60(0.707)(10^{-3}) \left[2(9.81)\left(\frac{132.6 - 8.42}{8.42}\right)(0.240) \right]^{1/2} = 3.54(10^{-3}) \text{ m}^3/\text{s}$$

$$V = \frac{Q_a}{A_0} = \frac{3.54(10^{-3})}{0.707(10^{-3})} = 5.01 \text{ m/s}$$

$$Re = \frac{\rho V d}{\mu} = \frac{858(5.01)(0.030)}{7.13(10^{-3})} = 18,087$$

From Fig. 12.29 at d/D = 0.30 and Re = 18,087: $\qquad C \cong 0.60$

Therefore: $\qquad Q_a = 3.54(10^{-3}) \text{ m}^3/\text{s}$

Exercise 12.19

From Table B-2: $\quad \gamma = 62.32 \text{ lb/ft}^3 \quad\quad \gamma_m = 844 \text{ lb/ft}^3$

From Eq. (12.37):

$$Q = CA \left[2g \left(\frac{p_o - p_i}{\gamma} + (z_o - z_i) \right) \right]^{1/2}$$

From the manometer equation (see Exercise 12.16):

$$\frac{\gamma_m - \gamma}{\gamma} h = \frac{p_o - p_i}{\gamma} + (z_o - z_i)$$

Therefore:
$$Q = CA \left[2g \left(\frac{\gamma_m - \gamma}{\gamma} \right) h \right]^{1/2}$$

$$A = (\pi/4)(D)^2 = \frac{\pi}{4} \left(\frac{6}{12} \right)^2 = 0.1963 \text{ ft}^2$$

$$C = 0.75 \text{ (Given)}$$

$$Q = CA \left[2g \left(\frac{\gamma_m - \gamma}{\gamma} \right) h \right]^{1/2}$$

$$= 0.75(0.1963) \left[2(32.2) \left(\frac{844 - 62.32}{62.32} \right) \left(\frac{h}{12} \right) \right]^{1/2} = 1.208\sqrt{h} \text{ ft}^3/s$$

$$Q = S_Q \sqrt{h} = 1.208\sqrt{h} \text{ ft}^3/s$$

Therefore: $\quad S_Q = 1.208 \text{ (ft}^3/s)/\sqrt{\text{in.Hg}}$

$$Q_{18.00} = 1.208\sqrt{18.00} = 5.125 \text{ ft}^3/s$$

$$Q_{18.05} = 1.208\sqrt{18.05} = 5.132 \text{ ft}^3/s$$

$$\Delta Q = Q_{18.05} - Q_{18.00} = 5.132 - 5.125 = 0.007 \text{ ft}^3/s$$

$$Q_{9.00} = 1.208\sqrt{9.00} = 3.624 \text{ ft}^3/s$$

$$Q_{9.05} = 1.208\sqrt{9.05} = 3.634 \text{ ft}^3/s$$

$$\mathcal{E} = \frac{Q_{9.05} - Q_{9.00}}{Q_{9.00}}(100) = \frac{3.634 - 3.624}{3.624}(100) = 0.276 \%$$

Exercise 12.20

From Eq. (12.38):

$$Q = CA_0 \sqrt{2g(h_1 - h_2)} = CA_0 \sqrt{2g\ \Delta h}$$
$$= C(\pi/4)(0.050)^2 \sqrt{2(9.81)(1.50)} = 10.65(10^{-3})\ C$$

Orifice Type	C	Flow Rate Q
Sharp Edged	0.61	0.00650 m^3/s
Rounded	0.98	0.01044 m^3/s
Short Tube	0.80	0.00852 m^3/s
Borda	0.51	0.00543 m^3/s

Exercise 12.21

From Eq. (12.38):

$$Q = CA_0 \sqrt{2g(h_1 - h_2)} = CA_0 \sqrt{2g\ \Delta h}$$
$$= C(\pi/4)(8/12)^2 \sqrt{2(32.2)(20)} = 12.53\ C\ \text{ft/s}^3$$

(a) With a rounded orifice: $C = 0.98$

$$Q = 12.53(0.98) = 12.28\ \text{ft}^3/\text{s}$$

(b) With a Borda orifice: $C = 0.51$

$$Q = 12.53(0.51) = 6.39\ \text{ft}^3/\text{s}$$

(c) Assumptions: $V_1 = V_2 = 0$

Exercise 12.22

From Eq. (12.40):
$$Q_a = \frac{C_v C_c A}{\sqrt{1 - (y_2/y_1)^2}} \sqrt{2g(y_1 - y_2)}$$

Given that: $C_c = 0.61$ and $C_v = 0.96$

Also: $y_2 = C_c d = 0.61(0.40) = 0.244$ m

$$Q_a = \frac{0.96(0.61)(3.0)(0.40)}{\sqrt{1 - (0.244/3.00)^2}} \sqrt{2(9.81)(3.00 - 0.244)} = 5.18 \text{ m}^3/\text{s}$$

$$V = \frac{Q_a}{A} = \frac{5.18}{100,000} = 0.0518(10^{-3}) \text{ m/s} = 0.0518 \text{ mm/s}$$

Exercise 12.23

A sketch of the weir is shown at the right:

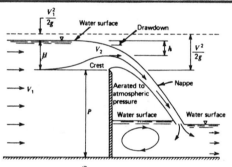

From Eq. (12.42):
$$H + \frac{V_1^2}{2g} = (H - h) + \frac{V_2^2}{2g}$$

From Eq. (12.43):
$$V_2 = \sqrt{2gh + V_1^2} \cong \sqrt{2gh}$$

Assumptions:
1. V_1 is small
2. No drawdown
3. V_2 is horizontal
4. End contraction is small

$$Q_i = \int_0^H V_2 L \, dh = \sqrt{2g} \, L \int_0^H \sqrt{h} \, dh = \frac{2}{3} \sqrt{2g} \, LH^{3/2}$$

$$Q_a = C_D Q_i = \frac{2}{3} \sqrt{2g} \, C_D LH^{3/2}$$

For English units with $C_D = 0.623$:

$$Q_a = \frac{2}{3} \sqrt{2g} \, C_D LH^{3/2} = \frac{2}{3} \sqrt{2(32.2)} (0.623) LH^{3/2} = 3.33 \, LH^{3/2}$$

For SI units with $C_D = 0.623$:

$$Q_a = \frac{2}{3} \sqrt{2g} \, C_D LH^{3/2} = \frac{2}{3} \sqrt{2(9.81)} (0.623) LH^{3/2} = 1.84 \, LH^{3/2}$$

Exercise 12.24

From the given equation for the coefficient of drag C_D:

$$C_D = 0.605 + 0.08 \frac{H}{P} + \frac{1}{305 H}$$

	Weir Coefficient C_D				
H	P = 0.2	P = 0.5	P = 1.0	P = 2.0	P = 5.0
0.10	0.678	0.624	0.616	0.612	0.609
0.20	0.701	0.653	0.637	0.629	0.625
0.40	0.773	0.677	0.645	0.629	0.620
0.60	—	0.706	0.658	0.634	0.620
0.80	—	0.737	0.673	0.641	0.622
1.00	—	—	0.688	0.648	0.624
1.20	—	—	0.704	0.656	0.627
1.40	—	—	0.719	0.663	0.630
1.60	—	—	0.735	0.671	0.633
1.80	—	—	0.751	0.679	0.636
2.00	—	—	—	0.687	0.639

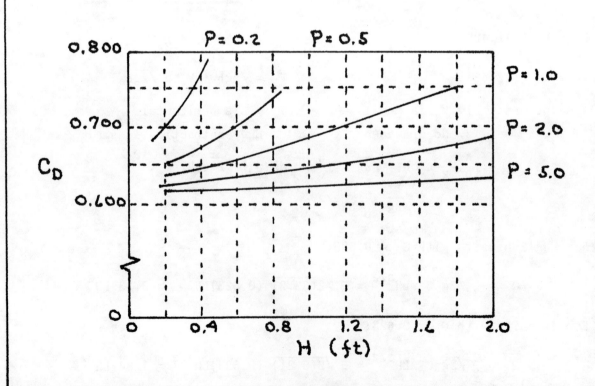

Exercise 12.25

From Eq. (12.46):
$$Q_a = \frac{2}{3}\sqrt{2g}\ C_D L H^{3/2}$$
$$= \frac{2}{3}\sqrt{2(9.81)}\ (0.62)\ LH^{3/2} = 1.83\ LH^{3/2}$$

Since the weir does not extend across the full width of the stream,
$$L_{eff} = L - 0.1nH = L - 0.1(2)H = L - 0.2H$$

For $H + P = 4.0$ m and $H/P < 0.4$: $\quad H < 1.143$ m. $\quad\quad$ Use $H = 1.0$ m

Since $Q_a = 1.83\ L_{eff} H^{3/2}$
$$L_{eff} = \frac{Q}{1.83 H^{3/2}} = \frac{6.0}{1.83(1.00)^{3/2}} = 3.28\ m$$

$L = L_{eff} + 0.2H = 3.28 + 0.2(1.00) = 3.48$ m $\quad\quad$ Use $L = 3.50$ m

With $H = 1.0$ m and $H + P = 4.0$ m: $\quad \frac{H}{P} = \frac{1}{3} = 0.33 < 0.40$

For the triangular weir: $\quad \theta = 90°$ $\quad\quad C_D \cong 0.59$

From Eq. (12.47): $\quad Q_a = \frac{8}{15}\sqrt{2g}\ \tan\frac{\theta}{2}\ C_D\ H^{5/2}$
$$= \frac{8}{15}\sqrt{2(9.81)}(\tan 45°)(0.59)\ H^{5/2} = 1.394\ H^{5/2}$$

For $Q_a = 6.0$ m³/s: $\quad H = \left[\frac{Q_a}{1.394}\right]^{2/5} = \left[\frac{6.0}{1.394}\right]^{2/5} = 1.793$ m

Therefore: $\quad P = 4.00 - 1.793 = 2.207$ m

For flow rate doubled:

For the rectangle: $\quad H = \left[\frac{Q_a}{1.83 L_{eff}}\right]^{2/3} = \left[\frac{12.0}{1.83(3.28)}\right]^{2/3} = 1.587$ m

$$\Delta H = 1.587 - 1.00 = 0.587\ m$$

For the triangle: $\quad H = \left[\frac{Q_a}{1.394}\right]^{2/5} = \left[\frac{12.0}{1.394}\right]^{2/5} = 2.366$ m

$$\Delta H = 2.366 - 1.793 = 0.573\ m$$

Slightly smaller change for the triangular 90-degree V-notch weir.

Exercise 12.26

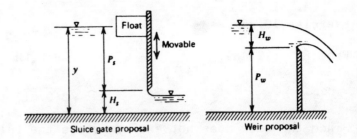

For the sluice gate:

From Eq. (12.41):

$$Q_a = C_{Ds} A \sqrt{2gy_1} = C_{Ds} LH_s \sqrt{2g(H_s + P_s)}$$

$$= C_{Ds} LH_s \sqrt{2g} \sqrt{P_s} \sqrt{1 + H_s/P_s}$$

With $H_s/P_s < 0.1$:

$$\sqrt{1 + H_s/P_s} = \sqrt{1 + (0.1)} = 1.049 < 1.05$$

Therefore:
$$Q_a = C_{Ds} \sqrt{2gP_s}\, LH_s$$

For the weir:

From Eq. (12.46):

$$Q_a = C_{Dw} \frac{2}{3} \sqrt{2g}\, LH_w^{3/2}$$

For the same flow rate:

$$Q_a = C_{Ds} \sqrt{2gP_s}\, LH_s = C_{Dw} \frac{2}{3} \sqrt{2g}\, LH_w^{3/2}$$

Therefore:
$$H_s = \frac{2C_{Dw} H_w^{3/2}}{3 C_{Ds} \sqrt{P_s}}$$

Since $C_{Dw} \cong C_{Ds}$:
$$P_s \cong y_0 \text{ (the desired fluid depth)}$$

$$H_s = \frac{2}{3} \sqrt{H_w/y_0})\, H_w \ll H_w$$

The sluice gate gives smaller fluid level variations but is more expensive to install and maintain. Select the weir unless the fluid level variations would be excessive.

Exercise 12.27

From Eq. (12.50):
$$\dot{m} = \frac{C_V C_E A_2 \rho_1}{\sqrt{1 - (A_2/A_1)^2}} \sqrt{2g \frac{P_1 - P_2}{\gamma_1}}$$

From Table B-4: $R = 1715$ ft·lb/(slug °R) $\quad k = 1.4$

$T = 150 + 460 = 610\,°R \quad A_1 = \infty$

$A_2 = (\pi/4)(1.00)^2 = 0.785$ in.^{2} = 0.00545 ft^2

$\gamma_1 = \dfrac{pg}{RT} = \dfrac{300(144)(32.2)}{1715(610)} = 1.3297$ lb/ft^3

$\rho_1 = \gamma_1/g = 1.3297/32.2 = 0.0413$ slug/ft^3

$p_2/p_1 = 250/300 = 0.833$

From Fig. 12.35: $C_E = 0.91$

Assume: $C_V = 0.98$

$\dot{m} = 0.98(0.91)(0.00545)(0.0413) \sqrt{2(32.2)\left(\dfrac{300 - 250}{1.3297}\right)(144)}$

$\quad = 0.1185$ slug/s $= 7.11$ slug/min.

From Eq. (12.51):

$$\left(\frac{P_2}{P_1}\right)_{critical} = \left(\frac{2}{k+1}\right)^{k/(k-1)} = \left(\frac{2}{1.4+1}\right)^{1.4/0.4} = 0.308$$

Therefore: $p_{min} = 0.308\, p_1 = 0.308(300) = 92.4$ psi

Exercise 12.28

From Eq. (12.50):
$$\dot{m} = \frac{C_V C_E A_2 \rho_1}{\sqrt{1 - (A_2/A_1)^2}} \sqrt{2g \frac{P_1 - P_2}{\gamma_1}}$$

From Table B-4: $R = 1554$ ft·lb/(slug °R) $\qquad k = 1.4$

$T = 70 + 460 = 530°R$

$A_1 = (\pi/4)(3.00)^2 = 7.069$ in.^{2} = 0.04909 ft^2

$A_2 = (\pi/4)(1.50)^2 = 1.767$ in.^{2} = 0.01227 ft^2

$\gamma_1 = \frac{pg}{RT} = \frac{150(144)(32.2)}{1554(530)} = 0.8445$ lb/ft^3

$\rho_1 = \gamma_1/g = 0.8445/32.2 = 0.02623$ slug/ft^3

$\sqrt{1 - (A_2/A_1)^2} = \sqrt{1 - (0.01227/0.04909)^2} = 0.968$

Assume: $C_V = 0.98 \qquad\qquad C_E = 0.85$

$\dot{m} = 6/32.2 = 0.1863$ slug/s

$$\frac{C_V C_E A_2 \rho_1}{\sqrt{1 - (A_2/A_1)^2}} = \frac{0.98(0.85)(0.01227)(0.02623)}{0.968} = 0.000277$$

$$\dot{m} = \frac{C_V C_E A_2 \rho_1}{\sqrt{1 - (A_2/A_1)^2}} \sqrt{2g \frac{P_1 - P_2}{\gamma_1}}$$

$$= 0.000277 \sqrt{2(32.2) \frac{P_1 - P_2}{0.8445}} = 0.1863$$

$$p_1 - p_2 = \frac{(0.1863)^2 (0.8445)}{(0.000277)^2 (2)(32.2)} = 5932 \text{ lb/ft}^2 = 41.2 \text{ psi}$$

Since $p_1 = 150$ psia: $\qquad p_2 = 150 - 41.2 = 108.8$ psia

$$\frac{p_2}{p_1} = \frac{108.8}{150} = 0.725$$

From Fig. 12.35: $\qquad C_E \cong 0.85$

Exercise 12.29

From Eq. (12.54):
$$Q = \frac{\pi D^4}{128\mu L}(\gamma_m - \gamma)h$$

From Table B-1:
$\gamma = 9.789 \text{ kN/m}^3 \qquad \gamma_m = 132.6 \text{ kN/m}^3$

$\mu = 1.002(10^{-3}) \text{ N} \cdot \text{s/m}^2 \qquad \rho = 998.2 \text{ kg/m}^3$

$$Q = \frac{\pi D^4}{128\mu L}(\gamma_m - \gamma)h$$

$$= \frac{\pi(0.001)^4(132.6 - 9.789)(10^3)(0.200)}{128(1.002)(10^{-3})(0.10)} = 6.02(10^{-6}) \text{ m}^3/\text{s}$$

At 25°C: $\mu = 0.890 \text{ N} \cdot \text{s/m}^2$

$$Q = 6.02(10^{-6})\frac{1.002}{0.890} = 6.78(10^{-6}) \text{ m}^3/\text{s}$$

$$\varepsilon = \frac{6.78 - 6.02}{6.02}(100) = 12.62 \%$$

Exercise 12.30

(a) Use L > 10 to 30 pipe diameters to straighten the flow.

(b)
$$W = \gamma Q t$$

Therefore $\qquad Qt = \frac{W}{\gamma} = \frac{200}{62.4} = 3.205 \text{ ft}^3$

For less than 1% error: $\qquad t_{min} = 100(0.05) = 5 \text{ s}$

$$Q_{max} = \frac{3.205}{5} = 0.641 \text{ ft}^3/\text{s}$$

(c) The time t to collect a certain volume of water (to calculate Q).
The manometer reading h.
The inlet and throat diameters (to check the coefficients).

(d) Plot Q versus h on log log paper.

$$Q_a = CA_0\sqrt{\frac{2g(\gamma_m - \gamma)}{\gamma}} = K\sqrt{h}$$

Thus, $\qquad \log Q_a = \log K + \frac{1}{2}\log h$

which is a straight line with a slope of $\frac{1}{2}$.

Exercise 12.31

A hot-film sensor can be employed to measure mass flow rate $\dot{m}$. The mass flow rate is given by the equation $\dot{m} = A\rho V$. The hot film responds to the heat transfer rate q (given by the equation $q = i^2 R$) which is related to ρV. The hot-film sensor is heated by the current i to a temperature above that of the fluid. The fluid then transports heat away from the film in proportion to the flow rate. The temperature difference $(\theta_a - \theta_f)$ in Eq. (12.55) is measured and a microprocessor is used to solve Eq. (12.55) to give $\dot{m} = A\rho V$ where A is the area of the venturi throat shown in Fig. 12.39.

Exercise 12.32

From Eq. (12.56): $\quad f_D = \dfrac{fV}{c}$

For a helium-neon laser: $\quad \lambda = 632.8$ nm

$\quad f = 4.7(10^{14})$ Hz

Also: $\quad c = 2.98(10^8)$ m/s

Therefore:

$$f_D = \frac{fV}{c} = \frac{4.7(10^{14})}{2.98(10^8)} = 1.577(10^6)\ V$$

Exercise 13.1

List of pressure readings in order of increasing magnitude (psi):

   81, 85, 87, 91, 92, 93, 94, 95, 96, 97, 97, 98,
   100, 100, 101, 101, 103, 103, 103, 105, 105, 107,
   109, 110, 112, 113, 113, 117, 118, 120, 123, 128

Relative frequency and cumulative frequency data:

Pressure	Number	Relative	Cumulative
81 - 90	3	0.09375	0.09375
91 - 100	11	0.34375	0.43750
101 - 110	10	0.31250	0.75000
111 - 120	6	0.18750	0.93750
121 - 130	2	0.06250	1.00000

The median is the average of the 16th and 17th readings. Thus:

$$\text{Median} = \frac{1}{2}(101 + 103) = 102 \text{ psi}$$

Exercise 13.2

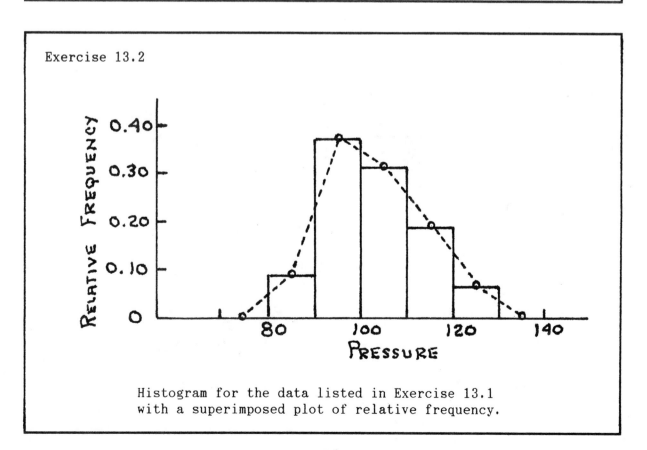

Histogram for the data listed in Exercise 13.1 with a superimposed plot of relative frequency.

Exercise 13.3

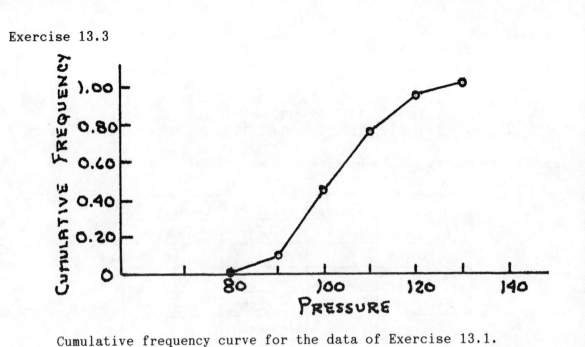

Cumulative frequency curve for the data of Exercise 13.1.

Exercise 13.4

For the data listed in Exercise 13.1:

From Eq. (13.1):
$$\bar{x} = \sum_{i=1}^{n} \frac{x_i}{n}$$

(a)

Column	$\sum x_i$	n	$\bar{x}$
1	833	8	104.125
2	861	8	107.625
3	727	8	90.875
4	876	8	109.500

(b) For columns 1 + 2: $\bar{x} = \dfrac{833 + 861}{8 + 8} = 105.875$

For columns 3 + 4: $\bar{x} = \dfrac{727 + 876}{8 + 8} = 100.1875$

(c) For all data: $\bar{x} = \dfrac{833 + 861 + 727 + 876}{8 + 8 + 8 + 8} = 103.03$

(d) With equal numbers of samples in each group, the average of the group means is the mean of all data.

ENGINEERING MEASUREMENTS by J. W. DALLY, W. F. RILEY, AND K. G. McCONNELL

Exercise 13.5

$$\text{Mode} = 95 \text{ psi}$$

$$\text{Median} = 102 \text{ psi}$$

$$\text{Mean} = 103.03 \text{ psi}$$

Exercise 13.6

List of tensile strength readings in order of increasing magnitude (psi):

820, 843, 847, 860, 862, 868, 871, 884, 887, 892, 897, 907, 913, 920, 922, 929, 933, 934, 936, 939, 942, 944, 950, 963, 968, 969, 971, 972, 977, 980, 981, 982, 983, 987, 990, 992, 999, 1013, 1021, 1031, 1043, 1043, 1066, 1072, 1079, 1080, 1091, 1115

Relative frequency and cumulative frequency data:

Strength	Number	Relative	Cumulative
800 - 849	3	0.0625	0.0625
850 - 899	8	0.1667	0.2292
900 - 949	11	0.2292	0.4583
950 - 999	15	0.3125	0.7708
1000 - 1049	5	0.1042	0.8750
1050 - 1099	5	0.1042	0.9792
1100 - 1149	1	0.0208	1.0000

The median is the average of the 24th and 25th readings. Thus:

$$\text{Median} = \frac{1}{2}(963 + 968) = 965.5 \text{ psi}$$

Exercise 13.7

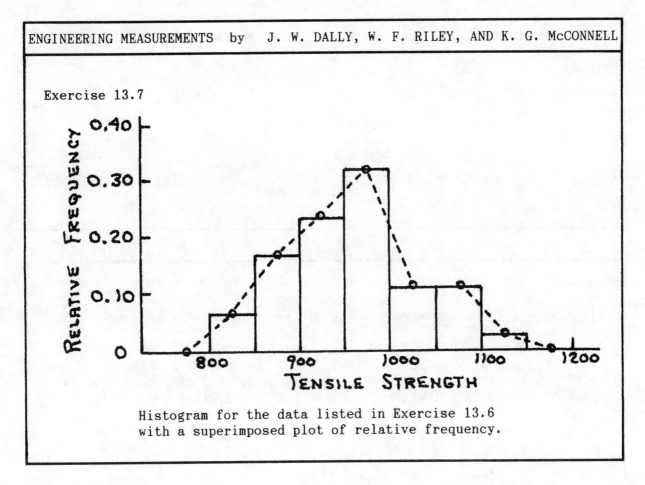

Histogram for the data listed in Exercise 13.6 with a superimposed plot of relative frequency.

Exercise 13.8

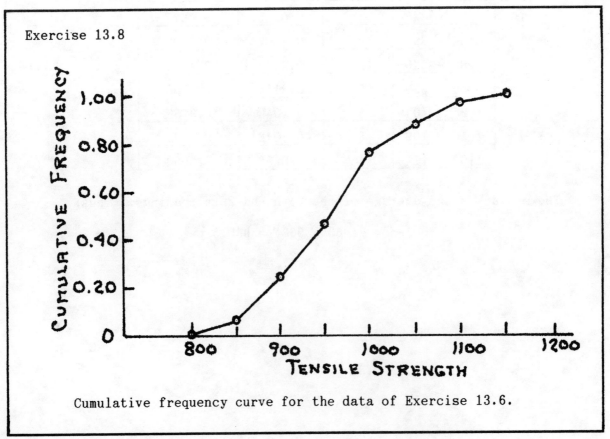

Cumulative frequency curve for the data of Exercise 13.6.

ENGINEERING MEASUREMENTS by J. W. DALLY, W. F. RILEY, AND K. G. McCONNELL

Exercise 13.9

For the data listed in Exercise 13.6:

From Eq. (13.1):
$$\bar{x} = \sum_{i=1}^{n} \frac{x_i}{n}$$

(a)

Column	$\sum x_i$	n	$\bar{x}$
1	7304	8	913.00
2	8005	8	1000.63
3	7976	8	997.00
4	7554	8	944.25
5	7037	8	879.63
6	8192	8	1024.00

(b) For columns 1 + 2: $\bar{x} = \dfrac{7304 + 8005}{8 + 8} = 956.81$

For columns 3 + 4: $\bar{x} = \dfrac{7976 + 7554}{8 + 8} = 970.63$

For columns 5 + 6: $\bar{x} = \dfrac{7037 + 8192}{8 + 8} = 951.81$

(c) For all data: $\bar{x} = \dfrac{46068}{48} = 959.75$

(d) With equal numbers of samples in each group, the average of the group means is the mean of all data.

Exercise 13.10

Mode = 975 psi

Median = 965.5 psi

Mean = 959.75 psi

Exercise 13.11

List of volume readings in order of increasing magnitude (gal):

    99.35, 99.42, 99.57, 99.68, 99.68, 99.68, 99.68, 99.68, 99.69,
    99.71, 99.72, 99.75, 99.77, 99.79, 99.81, 99.83, 99.84, 99.85,
    99.85, 99.85, 99.86, 99.86, 99.87, 99.87, 99.90, 99.91, 99.92,
    99.93, 99.95, 99.97, 99.97, 99.98, 99.99, 99.99, 100.02, 100.02,
    100.04, 100.04, 100.05, 100.06, 100.07, 100.08, 100.09, 100.12,
    100.16, 100.17, 100.18, 100.19, 100.21, 100.21, 100.27, 100.28,
    100.28, 100.29, 100.37, 100.37, 100.47, 100.67, 100.68, 100.89,

Relative frequency and cumulative frequency data:

Volume	Number	Relative	Cumulative
99.30 - 99.49	2	0.0333	0.0333
99.50 - 99.69	7	0.1167	0.1500
99.70 - 99.89	15	0.2500	0.4000
99.90 - 100.09	19	0.3167	0.7167
100.10 - 100.29	11	0.1833	0.9000
100.30 - 100.49	3	0.0500	0.9500
100.50 - 100.69	2	0.0333	0.9833
100.70 - 100.89	1	0.0167	1.0000

The median is the average of the 30th and 31st readings. Thus:

$$\text{Median} = \frac{1}{2}(99.97 + 99.97) = 99.97 \text{ gal}$$

The range of the flow meter is from 99.35 to 100.89 gal (% < 1%). The meter reads low 34/60 = 56.6 % of the time.

Exercise 13.12

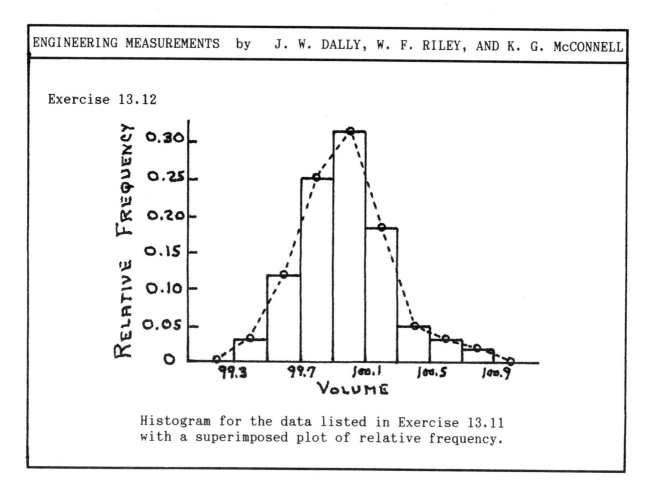

Histogram for the data listed in Exercise 13.11 with a superimposed plot of relative frequency.

Exercise 13.13

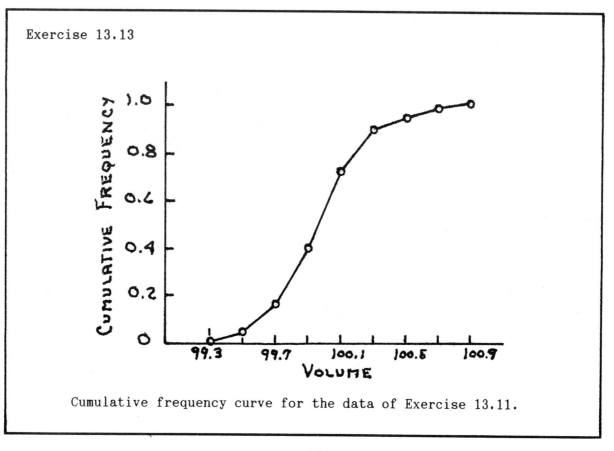

Cumulative frequency curve for the data of Exercise 13.11.

ENGINEERING MEASUREMENTS by J. W. DALLY, W. F. RILEY, AND K. G. McCONNELL

Exercise 13.14

For the data listed in Exercise 13.6:

From Eq. (13.1):
$$\bar{x} = \sum_{i=1}^{n} \frac{x_i}{n}$$

(a)

Column	$\sum x_i$	n	$\bar{x}$
1	1001.85	10	100.185
2	998.37	10	99.837
3	1000.57	10	100.057
4	998.64	10	99.864
5	1000.68	10	100.068
6	999.34	10	99.934

(b) For columns 1 + 2:
$$\bar{x} = \frac{1001.85 + 998.37}{10 + 10} = 100.01$$

For columns 3 + 4:
$$\bar{x} = \frac{1000.57 + 998.64}{10 + 10} = 99.96$$

For columns 5 + 6:
$$\bar{x} = \frac{1000.68 + 999.34}{10 + 10} = 100.00$$

(c) For columns 1 + 2 + 3:
$$\bar{x} = \frac{1001.85 + 998.37 + 1000.57}{10 + 10 + 10} = 100.03$$

For columns 4 + 5 + 6:
$$\bar{x} = \frac{998.64 + 1000.68 + 999.34}{10 + 10 + 10} = 99.96$$

(d) For all data:
$$\bar{x} = \frac{5999.45}{60} = 99.99$$

(e) With equal numbers of samples in each group, the average of the group means is the mean of all data.

Exercise 13.15

Mode = 100.00 gal

Median = 99.97 gal

Mean = 99.99 gal

Exercise 13.16

List of pressure readings in order of increasing magnitude (psi):

$$81, 85, 87, 91, 92, 93, 94, 95, 96, 97, 97, 98,$$
$$100, 100, 101, 101, 103, 103, 103, 105, 105, 107,$$
$$109, 110, 112, 113, 113, 117, 118, 120, 123, 128$$

$$\Sigma x_i = 3297$$

$$\Sigma (x_i - \bar{x})^2 = 3870.9663$$

$$\Sigma |x_i - \bar{x})| = 281.1875$$

From Eq. (13.1): $\bar{x} = \dfrac{1}{n} \Sigma x_i = \dfrac{1}{32}(3297) = 103.03125$

From Eq. (13.2): $S_x = \left[\sum_{i=1}^{n} \dfrac{(x_i - \bar{x})^2}{n-1} \right]^{1/2} = \left[\dfrac{3870.9663}{32-1} \right]^{1/2} = 11.1745$

$\sigma_n = \left[\sum_{i=1}^{n} \dfrac{(x_i - \bar{x})^2}{n} \right]^{1/2} = \left[\dfrac{3870.9663}{32} \right]^{1/2} = 10.9985$

From Eq. (13.3): $R = x_L - x_S = 128 - 81 = 47$

From Eq. (13.4): $d_x = \dfrac{\sum_{i=1}^{n} x_i - \bar{x}}{n} = \dfrac{281.1875}{32} = 8.7871$

From Eq. (13.5): $S_x^2 = \dfrac{\sum_{i=1}^{n} (x_i - \bar{x})^2}{n-1} = (11.1745)^2 = 124.8694$

From Eq. (13.6): $C_v = \dfrac{S_x}{\bar{x}}(100) = \dfrac{11.1745}{103.03125}(100) = 10.846$

Exercise 13.17

List of tensile strength readings in order of increasing magnitude (psi):

820, 843, 847, 860, 862, 868, 871, 884, 887, 892, 897, 907, 913, 920, 922, 929, 933, 934, 936, 939, 942, 944, 950, 963, 968, 969, 971, 972, 977, 980, 981, 982, 983, 987, 990, 992, 999, 1013, 1021, 1031, 1043, 1043, 1066, 1072, 1079, 1080, 1091, 1115

$$\Sigma x_i = 46,068$$
$$\Sigma (x_i - \bar{x})^2 = 239,479$$
$$\Sigma |x_i - \bar{x})| = 2748.50$$

From Eq. (13.1): $\quad \bar{x} = \dfrac{1}{n} \Sigma x_i = \dfrac{1}{48}(46,068) = 959.75$

From Eq. (13.2): $\quad S_x = \left[\displaystyle\sum_{i=1}^{n} \dfrac{(x_i - \bar{x})^2}{n-1} \right]^{1/2} = \left[\dfrac{239,479}{48 - 1} \right]^{1/2} = 71.38136$

$\qquad\qquad\qquad\quad \sigma_n = \left[\displaystyle\sum_{i=1}^{n} \dfrac{(x_i - \bar{x})^2}{n} \right]^{1/2} = \left[\dfrac{239,479}{48} \right]^{1/2} = 70.63389$

From Eq. (13.3): $\quad R = x_L - x_S = 1115 - 820 = 295$

From Eq. (13.4): $\quad d_x = \dfrac{\displaystyle\sum_{i=1}^{n} x_i - \bar{x}}{n} = \dfrac{2748.5}{48} = 8.7871$

From Eq. (13.5): $\quad S_x^2 = \dfrac{\displaystyle\sum_{i=1}^{n} (x_i - \bar{x})^2}{n-1} = (71.38136)^2 = 5095.30$

From Eq. (13.6): $\quad C_v = \dfrac{S_x}{\bar{x}}(100) = \dfrac{71.38136}{959.75}(100) = 7.4375$

ENGINEERING MEASUREMENTS by J. W. DALLY, W. F. RILEY, AND K. G. McCONNELL

Exercise 13.18

List of volume readings in order of increasing magnitude (gal):

   99.35, 99.42, 99.57, 99.68, 99.68, 99.68, 99.68, 99.68, 99.69,
   99.71, 99.72, 99.75, 99.77, 99.79, 99.81, 99.83, 99.84, 99.85,
   99.85, 99.85, 99.86, 99.86, 99.87, 99.87, 99.90, 99.91, 99.92,
   99.93, 99.95, 99.97, 99.97, 99.98, 99.99, 99.99, 100.02, 100.02,
   100.04, 100.04, 100.05, 100.06, 100.07, 100.08, 100.09, 100.12,
   100.16, 100.17, 100.18, 100.19, 100.21, 100.21, 100.27, 100.28,
   100.28, 100.29, 100.37, 100.37, 100.47, 100.67, 100.68, 100.89,

$$\Sigma x_i = 5999.45$$
$$\Sigma (x_i - \bar{x})^2 = 4.90124$$
$$\Sigma |x_i - \bar{x})| = 13.0364$$

From Eq. (13.1): $\bar{x} = \frac{1}{n} \Sigma x_i = \frac{1}{60}(5999.45) = 99.9908$

From Eq. (13.2): $S_x = \left[\sum_{i=1}^{n} \frac{(x_i - \bar{x})^2}{n-1} \right]^{1/2} = \left[\frac{4.90124}{60-1} \right]^{1/2} = 0.28822$

$\sigma_n = \left[\sum_{i=1}^{n} \frac{(x_i - \bar{x})^2}{n} \right]^{1/2} = \left[\frac{4.90124}{60} \right]^{1/2} = 0.28581$

From Eq. (13.3): $R = x_L - x_S = 100.89 - 99.35 = 1.54$

From Eq. (13.4): $d_x = \frac{\sum_{i=1}^{n} x_i - \bar{x}}{n} = \frac{13.0364}{60} = 0.21727$

From Eq. (13.5): $S_x^2 = \frac{\sum_{i=1}^{n}(x_i - \bar{x})^2}{n-1} = (0.28822)^2 = 0.08307$

From Eq. (13.6): $C_v = \frac{S_x}{\bar{x}}(100) = \frac{0.28822}{99.9908}(100) = 0.28825$

Exercise 13.19

For $\bar{x} = 80$ and $S_x = 3$:

Limits	z_1, z_2	$A(z_1, 0)$	$A(0, z_2)$	$p(z_1, z_2)$
73, 83	-2.333, +1	0.4902	0.3413	0.8315
70, 86	-3.333, +2	0.4988	0.4772	0.9760
67, 89	-4.333, +3	0.4999	0.4987	0.9986
80, 82	0, +0.667	0.0000	0.2475	0.2475
79, 80	-0.333, 0	0.1306	0.0000	0.1306
$-\infty$, 77	$-\infty$, -1.00	0.0000	0.1587	0.1587
86, ∞	2, ∞	0.0228	0.0000	0.0228

(a) 83.15%   (b) 97.60%   (c) 99.86%   (d) 24.75%

(e) 13.06%   (f) 15.87%   (g) 2.28%

Exercise 13.20

For $\bar{x} = 7.00$ and $S_x = 0.015$:

Limits	z_1, z_2	$A(z_1, 0)$	$A(0, z_2)$	$p(z_1, z_2)$
7.04, ∞	+2.667, +∞	0.0038	0.0000	0.0038
7.03, ∞	+2, +∞	0.0228	0.0000	0.0228
$-\infty$, 7.00	$-\infty$, 0	0.5000	0.0000	0.5000
$-\infty$, 6.94	$-\infty$, -4	0.0000	0.0001	0.0001
7.01, 7.03	+0.667, +2	0.2475	0.4772	0.2297
6.96, 6.99	-2.667, -0.667	0.4962	0.2475	0.2487
6.97, 7.03	-2, +2	0.4772	0.4772	0.9544

(a) 0.38%   (b) 2.28%   (c) 50.00%   (d) 0.01%

(e) 22.97%   (f) 24.87%   (g) 95.44%

Exercise 13.21

From Problem 13.11: $\bar{x} = 99.991 \quad S_x = 0.28822$

(a) $z_1 = \dfrac{1.01\,\bar{x} - \bar{x}}{S_x} = \dfrac{0.01(99.991)}{0.28822} = 3.4693$

(b) $z_1 = \dfrac{1.005\,\bar{x} - \bar{x}}{S_x} = \dfrac{0.005(99.991)}{0.28822} = 1.7346$

(c) $z_1 = \dfrac{1.002\,\bar{x} - \bar{x}}{S_x} = \dfrac{0.002(99.991)}{0.28822} = 0.6939$

(d) $z_1 = \dfrac{1.001\,\bar{x} - \bar{x}}{S_x} = \dfrac{0.001(99.991)}{0.28822} = 0.3469$

(e) $z_1 = \dfrac{0.99\,\bar{x} - \bar{x}}{S_x} = \dfrac{-0.01(99.991)}{0.28822} = -3.4693$

(f) $z_1 = \dfrac{0.995\,\bar{x} - \bar{x}}{S_x} = \dfrac{-0.005(99.991)}{0.28822} = -1.7346$

(f) $z_1 = \dfrac{0.998\,\bar{x} - \bar{x}}{S_x} = \dfrac{-0.002(99.991)}{0.28822} = -0.6939$

(f) $z_1 = \dfrac{0.999\,\bar{x} - \bar{x}}{S_x} = \dfrac{-0.001(99.991)}{0.28822} = -0.3469$

z_1	z_2	$A(z_1, 0)$	$A(0, z_2)$	$p(z_1, z_2)$
3.4693	$+\infty$	>0.4990	0.5000	<0.0010
1.7346	$+\infty$	0.4586	0.5000	0.0414
0.6939	$+\infty$	0.2561	0.5000	0.2439
0.3469	$+\infty$	0.1356	0.5000	0.3644
$-\infty$	-3.4693	0.5000	>0.4990	<0.0010
$-\infty$	-1.7346	0.5000	0.4586	0.0414
$-\infty$	-0.6939	0.5000	0.2561	0.2439
$-\infty$	-0.3469	0.5000	0.1356	0.3644

(a) <0.10% (b) 4.14% (c) 24.39% (d) 36.44%

(e) <0.10% (f) 4.14% (g) 24.39% (h) 36.44%

Exercise 13.22

From Eq. (13.11):
$$P(x) = 1 - e^{-[(x - x_0)/b]^m} \quad \text{for } x > x_0$$
$$P(x) = 0 \quad \text{for } x < x_0$$

Plots for $x_0 = 2$, $b = 5$, and $m = 1, 2, 5,$ and 10 are shown below.

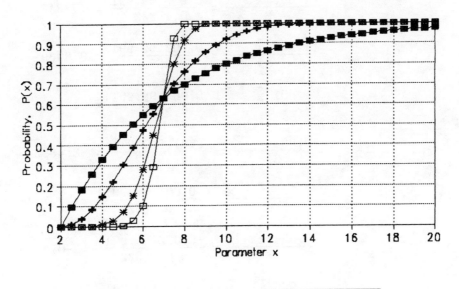

Exercise 13.23

For $m = 5$: $\quad P(2.5) = 0$

Exercise 13.24

For $m = 10$: $\quad P(3.0) = 0$

Exercise 13.25

For $m = 1$: $\quad P(4.0) = 0.33$

Exercise 13.26

From Eq. (13.17):
$$S_{\bar{x}} = \frac{S_x}{\sqrt{n}}$$

(a) For the data of Problem 13.1:

$$S_{\bar{x}} = 11.1745/\sqrt{32} = 1.9754$$

(b) For the data of Problem 13.6:

$$S_{\bar{x}} = 71.381/\sqrt{48} = 10.303$$

(c) For the data of Problem 13.11:

$$S_{\bar{x}} = 0.28822/\sqrt{60} = 0.03721$$

Exercise 13.27

From Eq. (13.18):
$$(\bar{x} - zS_{\bar{x}}) < \mu < (\bar{x} + zS_{\bar{x}})$$

From Problems 13.4, 13.16, and 13.26:

$$\bar{x} = 103.03 \qquad S_x = 11.1745 \qquad S_{\bar{x}} = 1.9754$$

Level	Width	Interval
99%	$\pm 2.57 S_{\bar{x}}$	$97.95 \leq \mu \leq 108.11$
95%	$\pm 1.96 S_{\bar{x}}$	$99.16 \leq \mu \leq 106.90$
90%	$\pm 1.65 S_{\bar{x}}$	$99.77 \leq \mu \leq 106.29$
80%	$\pm 1.28 S_{\bar{x}}$	$100.50 \leq \mu \leq 105.56$

ENGINEERING MEASUREMENTS by J. W. DALLY, W. F. RILEY, AND K. G. McCONNELL

Exercise 13.28

From Eq. (13.18): $(\bar{x} - zS_{\bar{x}}) < \mu < (\bar{x} + zS_{\bar{x}})$

From Problems 13.9, 13.17, and 13.26:

$\bar{x} = 959.75 \qquad S_x = 71.381 \qquad S_{\bar{x}} = 10.303$

Level	Width	Interval
99%	$\pm 2.57 S_{\bar{x}}$	$933.27 \le \mu \le 986.23$
95%	$\pm 1.96 S_{\bar{x}}$	$939.56 \le \mu \le 979.94$
90%	$\pm 1.65 S_{\bar{x}}$	$942.75 \le \mu \le 976.75$
80%	$\pm 1.28 S_{\bar{x}}$	$946.56 \le \mu \le 972.94$

Exercise 13.29

From Eq. (13.18): $(\bar{x} - zS_{\bar{x}}) < \mu < (\bar{x} + zS_{\bar{x}})$

From Problems 13.14, 13.18, and 13.26:

$\bar{x} = 99.9908 \qquad S_x = 0.28822 \qquad S_{\bar{x}} = 0.03721$

Level	Width	Interval
99%	$\pm 2.57 S_{\bar{x}}$	$99.895 \le \mu \le 100.086$
95%	$\pm 1.96 S_{\bar{x}}$	$99.918 \le \mu \le 100.064$
90%	$\pm 1.65 S_{\bar{x}}$	$99.929 \le \mu \le 100.052$
80%	$\pm 1.28 S_{\bar{x}}$	$99.943 \le \mu \le 100.038$

Exercise 13.30

List of measurements:

0.045	0.052	0.055	0.049	0.051
0.050	0.047	0.053	0.044	0.049

$$\Sigma x_i = 0.495$$

(a) From Eq. (13.1):

$$\bar{x} = \frac{1}{10} \Sigma x_i = \frac{1}{10}(0.495) = 0.0495$$

$$\Sigma(x_i - \bar{x})^2 = 0.0001085$$

(b) From Eq. (13.2):

$$S_x = \left[\sum_{i=1}^{n} \frac{(x_i - \bar{x})^2}{n-1}\right]^{1/2} = \left[\frac{0.0001085}{10-1}\right]^{1/2} = 0.003472$$

(c) From Eq. (13.17): $S_{\bar{x}} = S_x/\sqrt{n} = 0.003472/\sqrt{10} = 0.001098$

Since the sample is small (n < 20),

From Eq. (13.19): $\left[\bar{x} - t(\alpha)S_{\bar{x}}\right] < \mu < \left[\bar{x} + t(\alpha)S_{\bar{x}}\right]$

Level	α	ν	$t(\alpha)$	Confidence Interval
98%	0.990	9	2.82	$0.0464 \le \mu \le 0.0526$
95%	0.975	9	2.26	$0.0470 \le \mu \le 0.0520$
90%	0.950	9	1.83	$0.0475 \le \mu \le 0.0515$
80%	0.900	9	1.38	$0.0480 \le \mu \le 0.0510$

98% confident $C_D = 0.0495$ is within ±6.3%
95% confident $C_D = 0.0495$ is within ±5.0%
90% confident $C_D = 0.0495$ is within ±4.1%
80% confident $C_D = 0.0495$ is within ±3.1%

Exercise 13.31

From Exercise 13.30: $\bar{x} = 0.0495$ $S_x = 0.003472$

From Eq. (13.19): $\left[\bar{x} - t(\alpha)S_{\bar{x}} \right] < \mu < \left[\bar{x} + t(\alpha)S_{\bar{x}} \right]$

From Eq. (13.17): $S_{\bar{x}} = S_x / \sqrt{n}$

From Eq. (13.21): $n = \left[\dfrac{t(\alpha)S_x}{\delta} \right]^2$

For a 90% confidence level and an accuracy of 5% (±2.5%):

By trial and error with the data from Table 13.6:

$$\delta = 0.05(0.0495) = 0.002475$$

For $n = 10$, $\nu = 9$, $\alpha = 0.95$, and $t(\alpha) = 1.83$:

$$n = \left[\dfrac{t(\alpha)S_x}{\delta} \right]^2 = \left[\dfrac{1.83(0.003472)}{0.002475} \right]^2 = 6.6$$

For $n = 7$, $\nu = 6$, $\alpha = 0.95$, and $t(\alpha) = 1.94$:

$$n = \left[\dfrac{t(\alpha)S_x}{\delta} \right]^2 = \left[\dfrac{1.94(0.003472)}{0.002475} \right]^2 = 7.4$$

Therefore: Use 8 specimens

In a similar manner:

	Confidence Level			
δ	90 %	95 %	99 %	99.9 %
	n	n	n	n
0.10 $\bar{x}$	4	5	7	*
0.08 $\bar{x}$	5	6	9	*
0.05 $\bar{x}$	8	11	17	*
0.04 $\bar{x}$	11	15	25	*
0.02 $\bar{x}$	37	52	90	*
0.01 $\bar{x}$	140	198	345	*

*Specimen numbers for a confidence level of 99.9% can not be determined with the data listed in Table 13.6.

Exercise 13.32

List of measurements:

275°F	277°F	282°F	289°F	284°F
279°F	289°F	284°F	289°F	274°F

$$\Sigma x_i = 2822$$

(a) From Eq. (13.1):

$$\bar{x} = \frac{1}{10} \Sigma x_i = \frac{1}{10}(2822) = 282.2°F$$

$$\Sigma(x_i - \bar{x})^2 = 301.60$$

(b) From Eq. (13.2):

$$S_x = \left[\sum_{i=1}^{n} \frac{(x_i - \bar{x})^2}{n - 1}\right]^{1/2} = \left[\frac{301.60}{10 - 1}\right]^{1/2} = 5.7889$$

(c) From Eq. (13.17): $\quad S_{\bar{x}} = S_x/\sqrt{n} = 5.7889/\sqrt{10} = 1.8306$

Since the sample is small (n < 20),

From Eq. (13.19): $\quad \left[\bar{x} - t(\alpha)S_{\bar{x}}\right] < \mu < \left[\bar{x} + t(\alpha)S_{\bar{x}}\right]$

Level	α	ν	$t(\alpha)$	Confidence Interval
98%	0.990	9	2.82	$277.0 \leq \mu \leq 287.4$
95%	0.975	9	2.26	$278.1 \leq \mu \leq 286.3$
90%	0.950	9	1.83	$278.9 \leq \mu \leq 285.5$
80%	0.900	9	1.38	$279.7 \leq \mu \leq 284.7$

98% confident T = 282.2°F is within ±1.83%
95% confident T = 282.2°F is within ±1.47%
90% confident T = 282.2°F is within ±1.18%
80% confident T = 282.2°F is within ±0.89%

Exercise 13.33

From Exercise 13.32 $\bar{x} = 282.2°F \qquad S_x = 5.789°F$

From Eq. (13.19): $\left[\bar{x} - t(\alpha)S_{\bar{x}}\right] < \mu < \left[\bar{x} + t(\alpha)S_{\bar{x}}\right]$

From Eq. (13.17): $S_{\bar{x}} = S_x/\sqrt{n}$

From Eq. (13.21): $n = \left[\dfrac{t(\alpha)S_x}{\delta}\right]^2$

For a 90% confidence level and an accuracy of 5% (±2.5%):

By trial and error with the data from Table 13.6:

$$\delta = 0.05(282.2) = 14.11$$

For n = 4, ν = 3, α = 0.95, and t(α) = 2.35:

$$n = \left[\dfrac{t(\alpha)S_x}{\delta}\right]^2 = \left[\dfrac{2.35(5.789)}{14.11}\right]^2 = 0.93$$

For n = 2, ν = 1, α = 0.95, and t(α) = 6.31:

$$n = \left[\dfrac{t(\alpha)S_x}{\delta}\right]^2 = \left[\dfrac{6.31(5.789)}{14.11}\right]^2 = 6.70$$

Therefore: Use 3 specimens

In a similar manner:

δ	Confidence Level			
	90 %	95 %	99 %	99.9 %
	n	n	n	n
0.10 $\bar{x}$	2	3	4	*
0.08 $\bar{x}$	3	3	4	*
0.05 $\bar{x}$	3	4	5	*
0.04 $\bar{x}$	3	4	6	*
0.02 $\bar{x}$	5	7	11	*
0.01 $\bar{x}$	13	18	30	*

*Specimen numbers for a confidence level of 99.9% can not be determined with the data listed in Table 13.6.

Exercise 13.34

$$\bar{x} = 36{,}000 \text{ psi} \qquad S_x = 1250 \text{ psi} \qquad S_{min} = 32{,}000 \text{ psi}$$

$$z_2 = \frac{32{,}000 - 36{,}000}{1250} = -3.20$$

$$p(-\infty, -3.20) = 0.5000 - 0.4993 = 0.0007 = 0.07\%$$

The letter should specify the mean and standard deviation for the process and should indicate that there is a probability of 0.07% that any given aluminum rod will be under specification. To be able to specify that the yield strength for a specific production run is within ±1000 psi (±2.78%) at a confidence level of 99% it will be necessary to test the following number of specimens.

For a confidence level of 99%: $\qquad \alpha = 0.995$

From Eqs. (13.17) and (13.18):

$$\frac{t(\alpha)}{\sqrt{n}} = \frac{0.0278 \, \bar{x}}{S_x} = \frac{1000}{1250} = 0.80 \qquad \text{or} \qquad n = \left[\frac{t(\alpha)}{0.80}\right]^2$$

A trial and error solution using the data of Table 13.6 yields:

For $n = 10$, $\nu = 9$, and $t(\alpha) = 3.17 \qquad n = \left[\frac{3.25}{0.80}\right]^2 = 16.5$

For $n = 15$, $\nu = 14$, and $t(\alpha) = 2.98 \qquad n = \left[\frac{2.98}{0.80}\right]^2 = 13.9$

For $n = 14$, $\nu = 13$, and $t(\alpha) = 3.01 \qquad n = \left[\frac{3.01}{0.80}\right]^2 = 14.1$

Test 15 specimens for each production run.

Exercise 13.35

	Shipment A:	Shipment B:
	$n_A = 40$	$n_B = 60$
	$\bar{x}_d = 6.13$ mm	$\bar{x}_d = 6.06$ mm
	$\bar{x}_L = 25.4$ mm	$\bar{x}_L = 25.0$ mm
	$S_{xd} = 0.022$ mm	$S_{xd} = 0.034$ mm
	$S_{xL} = 0.140$ mm	$S_{xL} = 0.203$ mm

From Eq. (13.23):
$$S_p^2 = \frac{(n_1 - 1) S_{x1}^2 + (n_2 - 1) S_{x2}^2}{n_1 + n_2 - 2}$$

For the diameter:
$$S_p^2 = \frac{39(0.022)^2 + 59(0.034)^2}{40 + 60 - 2} = 0.0008886$$

For the length:
$$S_p^2 = \frac{39(0.140)^2 + 59(0.203)^2}{40 + 60 - 2} = 0.03261$$

From Eq. (13.22):
$$S_{(\bar{x}_2 - \bar{x}_1)}^2 = \frac{n_1 + n_2}{n_1 n_2} S_p^2$$

$$S_{(\bar{d}_2 - \bar{d}_1)}^2 = \frac{40 + 60}{40(60)} (0.0008886) = 0.00003703$$

$$S_{(\bar{L}_2 - \bar{L}_1)}^2 = \frac{40 + 60}{40(60)} (0.03261) = 0.001359$$

From Eq. (13.24):
$$t = \frac{|\bar{x}_2 - \bar{x}_1|}{S_{(\bar{x}_2 - \bar{x}_1)}}$$

For the diameter:
$$t = \frac{6.13 - 6.06}{\sqrt{0.00003703}} = 11.50$$

For the length:
$$t = \frac{25.4 - 25.0}{\sqrt{0.001359}} = 10.85$$

For $d = 98$ and $(1 - \alpha) = 0.05$ $t(\alpha) = 1.66$

For $d = 98$ and $(1 - \alpha) = 0.01$ $t(\alpha) = 2.37$

(a) Pins are not from the same batch.

(b) $t > t(\alpha)$; therefore, 99% confident the two shipments are different.

(c) Do not mix if the pins will be used in close tolerance situations.

ENGINEERING MEASUREMENTS by J. W. DALLY, W. F. RILEY, AND K. G. McCONNELL

Exercise 13.36

Shipment A:
- $n_A = 25$
- $\bar{x}_d = 6.05$ mm
- $\bar{x}_L = 24.9$ mm
- $S_{xd} = 0.02$ mm
- $S_{xL} = 0.20$ mm

Shipment B:
- $n_B = 15$
- $\bar{x}_d = 5.98$ mm
- $\bar{x}_L = 25.4$ mm
- $S_{xd} = 0.05$ mm
- $S_{xL} = 0.22$ mm

From Eq. (13.23):
$$S_p^2 = \frac{(n_1 - 1) S_{x1}^2 + (n_2 - 1) S_{x2}^2}{n_1 + n_2 - 2}$$

For the diameter:
$$S_p^2 = \frac{24(0.02)^2 + 14(0.05)^2}{25 + 15 - 2} = 0.0011737$$

For the length:
$$S_p^2 = \frac{24(0.20)^2 + 14(0.22)^2}{25 + 15 - 2} = 0.043095$$

From Eq. (13.22):
$$S_{(\bar{x}_2 - \bar{x}_1)}^2 = \frac{n_1 + n_2}{n_1 n_2} S_p^2$$

$$S_{(\bar{d}_2 - \bar{d}_1)}^2 = \frac{25 + 15}{25(15)} (0.0011737) = 0.0001252$$

$$S_{(\bar{L}_2 - \bar{L}_1)}^2 = \frac{25 + 15}{25(15)} (0.043095) = 0.004597$$

From Eq. (13.24):
$$t = \frac{|\bar{x}_2 - \bar{x}_1|}{S_{(\bar{x}_2 - \bar{x}_1)}}$$

For the diameter:
$$t = \frac{6.05 - 5.98}{\sqrt{0.0001252}} = 6.256$$

For the length:
$$t = \frac{24.9 - 25.4}{\sqrt{0.004597}} = 7.374$$

For $d = 38$ and $(1 - \alpha) = 0.05$ $t(\alpha) = 1.69$

For $d = 98$ and $(1 - \alpha) = 0.01$ $t(\alpha) = 2.43$

(a) Pins are not from the same batch.

(b) $t > t(\alpha)$; therefore, 99% confident the two shipments are different.

(c) Do not mix if the pins will be used in close tolerance situations.

Exercise 13.37

From Eq. (13.1): $\bar{x} = 763.77$ mm

From Eq. (13.2): $S_x = 0.4200$ mm

From Eq. (13.25): $DR = (x_i - \bar{x})/S_x$

Reading	x_i	$x_i - \bar{x}$	DR
1	763.2	-0.57	1.36
2	764.1	0.33	0.79
3	764.6	0.83	1.98
4	764.2	0.43	1.02
5	763.5	-0.27	0.64
6	763.5	-0.27	0.64
7	763.4	-0.37	0.88
8	764.3	0.53	1.26
9	764.1	0.33	0.79
10	764.0	0.23	0.55
11	763.3	-0.47	1.12
12	763.4	-0.37	0.88
13	763.7	-0.07	0.17
14	763.8	0.03	0.07
15	763.5	-0.27	0.64

From Table 13.7 for n = 15: $DR_0 = 2.13$

All $DR < DR_0$; therefore, no points are removed.

Exercise 13.38

Since no points were removed:

From Eq. (13.1): $\bar{x} = 763.77$ mm

From Eq. (13.2): $S_x = 0.4200$ mm

Exercise 13.39

From Eq. (13.17):
$$S_{\bar{x}} = S_x/\sqrt{n}$$

$$S_{\bar{x}}(\text{improved}) = S_x/\sqrt{N}$$

For an increase in precision by a factor of 1.2:

$$S_{\bar{x}}(\text{improved}) = \frac{S_{\bar{x}}}{1.2} = \frac{S_x}{\sqrt{N}} = \frac{S_x/\sqrt{n}}{1.2}$$

$$N = (1.2)^2 n = 1.44\, n$$

Additional measurements = $(1.44 - 1)n = 0.44n$

Exercise 13.40

From Problem 13.39:
$$N = k^2 n$$
$$\Delta = (k^2 - 1)n$$

Case	k	k^2	Δ
a	1.25	1.56	0.56
b	1.50	2.25	1.25
c	1.75	3.06	2.06
d	2.00	4.00	3.00
e	5.00	25.00	24.00
f	10.00	100.00	99.00

Exercise 13.41

For: $\mu = 1$ and $\sigma = 1$

$t(\alpha)S_x/\sqrt{n}$	$t(\alpha)/\sqrt{n}$	Confidence Level			
		90%	95%	99%	99.9%
		n	n	n	n
± 0.10μ	10	-	2	3	4
± 0.05μ	5	2	3	4	5
± 0.02μ	2	3	4	6	8
± 0.01μ	1	5	7	11	17
± 0.005μ	0.5	13	18	30	50
± 0.002μ	0.2	70	99	170	280
± 0.001μ	0.1	275	385	675	1100

Exercise 13.42

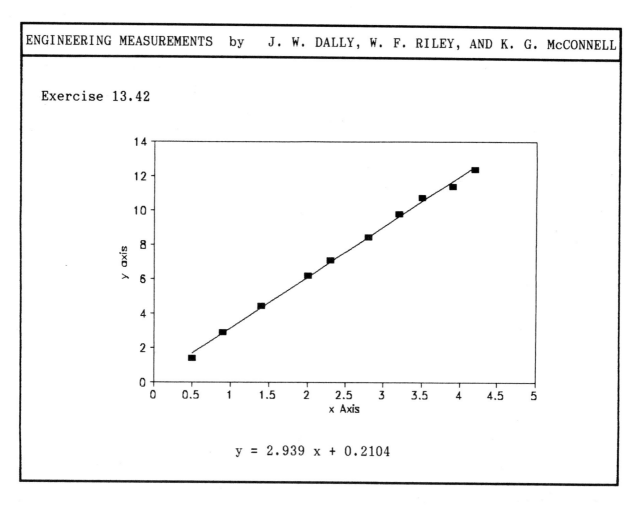

y = 2.939 x + 0.2104

Exercise 13.43

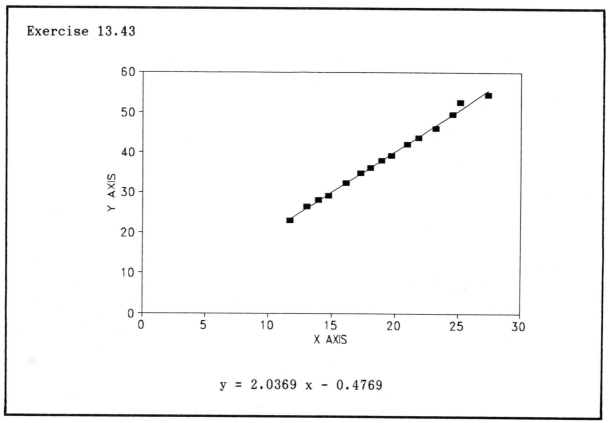

y = 2.0369 x − 0.4769

Exercise 13.44

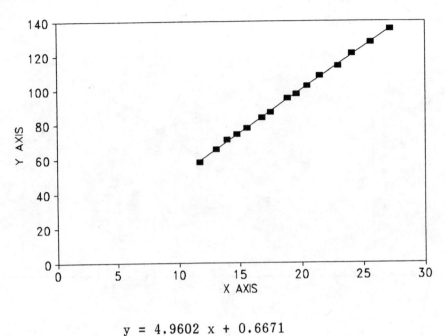

$$y = 4.9602\ x + 0.6671$$

Exercise 13.45

From Eq. (13.29):
$$R^2 = 1 - \frac{n-1}{n-2}\left[\frac{\{y^2\} - m\{xy\}}{\{y^2\}}\right]$$

where $\{y^2\} = \Sigma y^2 - (\Sigma y)^2/n$

$\{xy\} = \Sigma xy - (\Sigma x)(\Sigma y)/n$

For the data of Exercise 13.42: $R^2 = 0.998$

Exercise 13.46

For the data of Exercise 13.43: $R^2 = 0.996$

Exercise 13.47

For the data of Exercise 13.44: $R^2 = 0.999$

Exercise 13.48

$$C = a e^{-mt} \quad \text{or} \quad \ln C = \ln a - mt$$

t	C	ln C	ln C (est)	C (est)
10	2.05	0.7178	0.6781	1.970
20	2.00	0.6931	0.6715	1.957
30	1.96	0.6729	0.6648	1.944
40	1.92	0.6523	0.6581	1.931
50	1.89	0.6366	0.6515	1.918
100	1.79	0.5822	0.6181	1.855
200	1.68	0.5188	0.5514	1.736
500	1.45	0.3716	0.3514	1.421

$\ln a = 0.6848 \quad a = 1.983$

$m = -0.0006668$

$C = 1.983 \, e^{-0.0006668 \, t}$

Exercise 13.49

From Eq. (13.30): $\quad Y_i = a + b_1 x_1 + b_2 x_2 + b_3 x_3$

The matrix is singular when x_3 is considered. Note that x_3 varies periodically with increasing y.

For x_1 only: $\quad y = 0.4070 x_1 + 7.0071$

For x_1 and x_2: $\quad y = 0.6000 x_1 - 0.3067 x_2 + 7.3567$

Exercise 13.50

From Eqs. (13.29) and (13.34):

For x_1 only: $\quad R^2 = 0.9301$

For x_1 and x_2: $\quad R^2 = 0.9425$

Exercise 13.51

(a) From Eq. (13.1):
$$\bar{x} = \frac{1}{n} \Sigma x_i = \frac{1}{N} \Sigma n_i W_i$$

$$\bar{x} = \frac{1}{21,321} [22(30.5) + 480(30.7) + 5106(30.9) + 10,461(31.1)$$
$$+ 4618(31.3) + 619(31.5) + 15(31.7)] = 31.098 \text{ lb}$$

From Eq. (13.2):
$$S_x^2 = \sum_{i=1}^{n} \frac{(x_i - \bar{x})^2}{n-1} = \sum_{i=1}^{n} \frac{(W_i - \bar{x})^2 (n)}{N-1}$$

$$S_x^2 = \frac{1}{21,320} [(30.5 - 31.098)^2 (22) + (30.7 - 31.098)^2 (480)$$
$$+ (30.9 - 31.098)^2 (5106) + (31.1 - 31.098)^2 (10462)$$
$$+ (31.3 - 31.098)^2 (4618) + (31.5 - 31.098)^2 (619)$$
$$+ (31.7 - 31.098)^2 (15) = \frac{578.02}{21,320} = 0.0271116$$

Therefore:
$$S_x = 0.1647 \text{ lb}$$

From Eq. (13.8):
$$z = \frac{W_i - \bar{x}}{S_x}$$

Group Interval	z_1	z_2	$p(z_1, z_2)$	n_E	n_O	$(n_O - n_E)^2$
≤ 30.59	-∞	-3.024	0.0013	28	22	36
30.60 - 30.79	-3.024	-1.809	0.0338	721	480	58081
30.80 - 30.99	-1.809	-0.595	0.2401	5119	5106	169
31.00 - 31.19	-0.595	0.619	0.4562	9727	10461	536756
31.20 - 31.39	0.619	1.834	0.2346	5002	4618	147456
31.40 - 31.59	1.834	3.048	0.0321	684	619	65
≥ 31.60	3.048	∞	0.0012	26	15	121

(b) From Eq. (13.35):
$$\chi^2 = \sum \frac{(n_O - n_E)^2}{n_E} = 171.6$$

From Table 13.9 (n = 7, k = 2, therefore ν = 5): $\chi^2 = 15.1$ (99%)

Can conclude with 99% certainty that the distribution is not normal.

(c) Only if a relationship between quality and weight has been found by actual experience.

Exercise 13.52

3200 castings   Rejections: $n_E = 3200(0.04) = 128$ (8 hr)

800 castings $n_O = 22$ (2 hr) $= 88$ (8 hr)

From Eq. (13.35):
$$\chi^2 = \frac{(n_O - n_E)^2}{n_E} = \frac{(88 - 128)^2}{128} = 12.5$$

From Table 13.9 ($\nu = 1$): $X^2 = 6.63$ (99%)

Can conclude with 99% certainty that the new process reduces rejects and that the reduction in defects is not due to chance variations.

Exercise 13.53

Component	$S_{\bar{x}}$	$S_{\bar{x}}^2$
Shoulder	0.040	0.0016
Bearing	0.030	0.0009
Sleeve 1	0.080	0.0064
Gear	0.050	0.0025
Sleeve 2	0.070	0.0049
Nut	0.080	0.0064

$$\sum S_{\bar{x}}^2 = 0.0227$$

From Eq. (13.37):

$$S_{\bar{y}} = \sqrt{S_{\bar{x}1}^2 + S_{\bar{x}2}^2 + \cdots + S_{\bar{x}n}^2} = \sqrt{\sum S_{\bar{x}}^2} = \sqrt{0.0227} = 0.1507 \text{ mm}$$

(b) For a normal distribution:

$$p(-1,+1) = 0.6826 = 68.3\ \%$$

Exercise 13.54

$$V = \frac{4\pi}{3}\left(\frac{d}{2}\right)^3 = \frac{\pi}{6}d^3$$

Let:
$$\bar{x}_1 = \bar{x}_2 = \bar{x}_3 = d = 50 \text{ mm} \qquad \bar{x}_4 = \pi/6$$
$$S_{\bar{x}1} = S_{\bar{x}2} = S_{\bar{x}3} = 0.05 \text{ mm} \qquad S_{\bar{x}4} = 0$$

From Eq. (13.38):
$$S_{\bar{y}} = (\bar{x}_1 \bar{x}_2 \cdots \bar{x}_n)\sqrt{\frac{S_{\bar{x}1}^2}{\bar{x}_1^2} + \frac{S_{\bar{x}2}^2}{\bar{x}_2^2} + \cdots + \frac{S_{\bar{x}n}^2}{\bar{x}_n^2}}$$

$$= \frac{\pi}{6}d^3\sqrt{3S_{\bar{x}1}^2/\bar{x}^2} = \frac{\pi}{6}(50)^3\sqrt{3(0.05)^2/(50)^2}$$

$$= 113.4 \text{ mm}^3$$

Exercise 13.55

$$\sigma_x = \frac{E}{1-\nu^2}\left[\varepsilon_x + \nu\varepsilon_y\right]$$

$$d\sigma_x = \frac{E}{1-\nu^2}\left[d\varepsilon_x + \nu d\varepsilon_y + \varepsilon_y d\nu\right]$$

$$+ \left[\varepsilon_x + \nu\varepsilon_y\right]\frac{(1-\nu^2)dE - 2E\nu\, d\nu}{(1-\nu^2)^2}$$

$$\frac{d\sigma_x}{\sigma_x} = \frac{d\varepsilon_x + \nu d\varepsilon_y + \varepsilon_y d\nu}{\varepsilon_x + \nu\varepsilon_y} + \frac{(1-\nu^2)dE - 2E\nu\, d\nu}{E(1-\nu^2)}$$

$$= \frac{d\varepsilon_x}{\varepsilon_x + \nu\varepsilon_y} + \frac{\nu d\varepsilon_y}{\varepsilon_x + \nu\varepsilon_y} + \frac{\varepsilon_y d\nu}{\varepsilon_x + \nu\varepsilon_y} + \frac{dE}{E} + \frac{2\nu\, d\nu}{1-\nu^2}$$

For the case $\nu = \frac{1}{3}$ and $\varepsilon_x = \varepsilon_y = \varepsilon$:

$$\frac{d\sigma_x}{\sigma_x} = \frac{3}{4}\frac{d\varepsilon}{\varepsilon} + \frac{1}{4}\frac{d\varepsilon}{\varepsilon} + \frac{d\nu}{1+\nu} + \frac{dE}{E} + \frac{2\nu\, d\nu}{1-\nu^2}$$

$$= \frac{d\varepsilon}{\varepsilon} + \frac{dE}{E} + \frac{\nu}{1-\nu}\frac{d\nu}{\nu}$$

$$= 0.03 + 0.05 + \frac{1}{2}(0.05) = 0.105 = 10.5 \%$$